SpringerBriefs in Mathematics

Series Editors

Nicola Bellomo, Torino, Italy

Michele Benzi, Pisa, Italy

Palle Jorgensen, Iowa, USA

Tatsien Li, Shanghai, China

Roderick Melnik, Waterloo, Canada

Otmar Scherzer, Linz, Austria

Benjamin Steinberg, New York, USA

Lothar Reichel, Kent, USA

Yuri Tschinkel, New York, USA

George Yin, Detroit, USA

Ping Zhang, Kalamazoo, USA

SpringerBriefs in Mathematics showcases expositions in all areas of mathematics and applied mathematics. Manuscripts presenting new results or a single new result in a classical field, new field, or an emerging topic, applications, or bridges between new results and already published works, are encouraged. The series is intended for mathematicians and applied mathematicians. All works are peer-reviewed to meet the highest standards of scientific literature.

Titles from this series are indexed by Web of Science, Mathematical Reviews, and zbMATH.

More information about this series at http://www.springer.com/series/10030

Colton Magnant · Pouria Salehi Nowbandegani

Topics in Gallai-Ramsey Theory

Colton Magnant
Advanced Analytics Group
United Parcel Service
Atlanta, GA, USA

Pouria Salehi Nowbandegani
Department of Mathematics
Vanderbilt University
Nashville, TN, USA

ISSN 2191-8198   ISSN 2191-8201   (electronic)
SpringerBriefs in Mathematics
ISBN 978-3-030-48896-3   ISBN 978-3-030-48897-0   (eBook)
https://doi.org/10.1007/978-3-030-48897-0

Mathematics Subject Classification: 05C15, 05C55, 05C75, 05C76, 05C90, 05C05, 05C07, 05C10, 05C12, 05C35, 05C38, 05C40, 05C42, 05C45

This Springer imprint is published by the registered company Springer Nature Switzerland AG
The registered company address is: Gewerbestrasse 11, 6330 Cham, Switzerland

Preface

[1] ... Gould, M. ... Gallai–Ramsey number ... ...
... Ramsey number ... Comb. ... vol. ... 2016.
[2] ... Gould, ... K. Ohba, ... The dynamic survey of rainbow ... graph ...
... ... (2016).
[3] ... Magnant, P. Ohba, Rainbow generalizations of Ramsey theory — a dynamic
survey. Theor. App. Grad. (2020) ... .

The original work on Gallai-Ramsey numbers from this perspective [1] was initiated in meetings between Ralph Faudree, Ron Gould, Mike Jacobson, and Colton Magnant, on a visit to University of Memphis. In the comforts of the Provost's office, we considered forbidding several small rainbow subgraphs before arriving at the rainbow triangle as the most interesting candidate (only to later discover Gallai's result, Theorem 2.1).

Upon completing the proof of one of the results in [1], we proudly drew a little square on the board to indicate the end of the proof. Much in the style of Paul Erdös, Faudree immediately said "How about we consider...", but his question got cut off. Jacobson jokingly shouted: "No! We are going to cherish this moment for at least three minutes!" It is moments like this and many others that serve as a constant reminder of the profound and everlasting influence that Ralph Faudree had on his friends, collaborators, and the field of Graph Theory, particularly in Ramsey Theory.

The motivation for writing this text was originally to help graduate students dive into the exciting and ever-expanding area of rainbow generalizations of Ramsey theory. A secondary motivation grew out of this: to cultivate and inspire new research.

We truly hope that the explanations and compilation of results meet these objectives. For the reader interested in keeping up with current developments in the area, please see the dynamic survey [3] (originally published in [2]) for an updated listing of known results in the area.

Atlanta, USA   Colton Magnant
Nashville, USA   Pouria Salehi Nowbandegani

References

1. R. J. Faudree, R. Gould, M. Jacobson, C. Magnant, Ramsey numbers in rainbow triangle free colorings. Australas. J. Combin. **46**, 269–284, 2010
2. S. Fujita, C. Magnant, K. Ozeki, Rainbow generalizations of Ramsey theory: a survey. Graphs Combin. **26**(1), 1–30, 2010
3. S. Fujita, C. Magnant, K. Ozeki, Rainbow generalizations of Ramsey theory - a dynamic survey. Theo. Appl. Graphs **0**(1), 2014

Contents

Chapter 1
Introduction and Basic Definitions

A *graph* $G = (V, E)$ is a collection of vertices $V = V(G)$ and a collection of (unordered) pairs of vertices called edges $E = E(G)$. Since each edge $e = uv$ is a pair of vertices, we call each vertex u or v in e an *end* of the edge e. The two vertices at either end of an edge are *adjacent*. A graph is called *simple* if every pair of vertices is contained in at most one edge. A graph is called *finite* if it has a finite number of vertices and so, for a simple graph, also a finite number of edges. In this work, we consider only finite simple graphs so we simply use the term graph to mean a finite simple graph. The number of vertices in G, called the *order* of G, is commonly denoted by $n = |G| = |V(G)|$, and the number of edges, called the *size* of G, is commonly denoted by $m = |E(G)|$.

Two graphs G and H are called *isomorphic*, denoted by $G \cong H$, if there is a permutation π of the vertices of G such that $\pi(G) = H$, that is, u is adjacent to v (in G) if and only if $\pi(u)$ is adjacent to $\pi(v)$ (in H). Oftentimes notation is abused and isomorphic graphs are simply called equal. A *subgraph* of a graph G is a subset of vertices $U \subset V(G)$ and a corresponding subset of edges $F \subseteq E(G)$ such that the edges of F use only the vertices of U. A subgraph is *spanning* if $U = V$. A subgraph is *induced*, denoted by $G[U]$, if the subset of edges F contains all the edges of $E(G)$ which use precisely the vertices of U. Note that the only spanning induced subgraph of a graph G is G itself.

The topic of this text involves assigning colors to the edges of graphs so we say a graph G is *edge-colored* if there is an associated function $c : E(G) \rightarrow C$ where C is some set of colors. Oftentimes, if $|C| = k$, it is convenient to think of C as $C = [k] = \{1, 2, \dots, k\}$. When k is small, named colors are frequently used in place of numbers, e.g. red, blue and green. Unless specifically stated otherwise, we assume all colorings are colorings of the edges and so a graph will be called *colored* if its edges are assigned a coloring. For an edge $e = uv$, we sometimes use the notation $c(uv)$ to refer to the color assigned to the edge by the function c. A colored graph is called *rainbow* if all of its edges receive distinct colors and *monochromatic* if all of its edges receive a single color.

© The Author(s), under exclusive license to Springer Nature Switzerland AG 2020
C. Magnant and P. Salehi Nowbandegani, *Topics in Gallai-Ramsey Theory*,
SpringerBriefs in Mathematics, https://doi.org/10.1007/978-3-030-48897-0_1

Given a colored graph G, define G_i to be the spanning subgraph of G containing precisely the edges of color i. Also, for convenience, define $G_{\geq i}$ to be the union of the graphs G_j (all on the same vertex set) over the values of j with $j \geq i$. That is,

$$G_{\geq i} = \bigcup_{j \geq i} G_j.$$

1.1 Classes of Graphs

The bulk of the results in Ramsey Theory and Gallai-Ramsey Theory are sharp numbers and bounds on the numbers for specific classes of graphs. For this reason, we describe several classes of graphs in this section that will be used at various points in this text. These (or equivalent) definitions can also be found in standard texts like [1] by Chartrand et al. and many of these definitions are borrowed from [2] by Li et al. We include many of the definitions here for completeness.

For a fixed vertex v, the set of all vertices adjacent to v is the *neighborhood* of v and denoted by $N(v)$. The order of $N(v)$ (equivalently, the number of edges in $E(G)$ containing v) is called the *degree* of v and denoted by $deg(v)$ or $d(v)$. The smallest degree over all vertices in $V(G)$ is the *minimum degree*, denoted by $\delta(G)$, and the largest degree over all vertices in $V(G)$ is the *maximum degree*, denoted by $\Delta(G)$. When we restrict to only a subgraph $H \subseteq G$ (or when the graph under consideration is not clear from context), we denote by $d_H(x)$ and $N_H(x)$ the degree and the neighborhood of a vertex x within H, respectively.

For example, in the graph in Fig. 1.1, we have $d(v_1) = \Delta(G) = 4$, $d(v_2) = 2$, $d(v_3) = 3$, $d(v_4) = \delta(G) = 1$, $d(v_5) = \Delta(G) = 4$ and $d(v_6) = 2$. A vertex with degree 1 (like v_4) is called a *pendant vertex* or *end vertex* or, particularly within the context of trees, a *leaf*. A vertex with degree 0 is called an *isolated vertex*.

The unique graph of order n containing all possible edges, all $\binom{n}{2}$ of them, is the *complete graph* (or *clique*) and denoted by K_n. See Fig. 1.2 for an example of a complete graph. Given a graph G with a subgraph H, we call a subgraph of G that is isomorphic to H a *copy* of H in G. This term will be used frequently throughout this work since Ramsey Theory in general is the search for a monochromatic copy of a particular graph within a (larger) edge-colored (most often complete) graph.

Fig. 1.1 A graph
demonstrating degrees

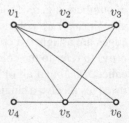

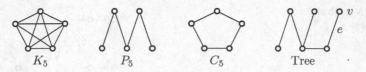

Fig. 1.2 A complete graph, a path, a cycle and a tree

Fig. 1.3 The graphs S_4^+, $S_4 = K_{1,4}$ and $K_{3,4}$

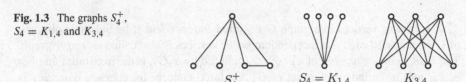

A *path* on n vertices, denoted by P_n, is the graph consisting of an ordered set of vertices $\{v_1, v_2, \ldots, v_n\}$ and all edges of the form $v_i v_{i+1}$ for $1 \le i \le n - 1$, that is, edges between consecutive vertices in the list. A *cycle* on n vertices, denoted by C_n, is the graph consisting of a path on n vertices as defined above, with the addition of the edge $v_1 v_n$. A *tree* is any connected graph containing no cycle. In particular, every path is a tree. See Fig. 1.2 for an example of a clique (K_5), a path (P_5), a cycle (C_5) and a more general tree.

A graph G is *bipartite* if there is a (bi)partition of $V(G)$ into two sets A and B so that for every edge $e \in E(G)$, one end of e is in A and the other end is in B. For positive integers a and b, a bipartite graph $G = A \cup B$ is called *complete bipartite*, denoted by $K_{a,b}$, if $|A| = a$ and $|B| = b$ and all edges from A to B are present in $E(G)$. In particular, all paths and trees are bipartite, all even cycles are bipartite while all odd cycles are not, and all complete graphs K_n for $n \ge 3$ are not bipartite. Equivalently, we can prove that a graph is bipartite if and only if it contains no odd cycle as a subgraph. The complete bipartite graph $K_{1,t}$ is also known as a *star* on $t + 1$ vertices and denoted by S_t. Note that a star is also a tree and all trees are bipartite. See Fig. 1.3 for two examples of complete bipartite graphs, namely, $S_4 = K_{1,4}$ and $K_{3,4}$.

Given two graphs G and H, the *union* of G and H, denoted by $G \cup H$, is the graph with vertex set consisting of the disjoint union $V(G) \cup V(H)$ and edge set consisting of the disjoint union $E(G) \cup E(H)$. When taking the union of $k \ge 2$ copies of a single graph G, we use the notation kG in place of $G \cup G \cup \cdots \cup G$. The *join* of G and H, denoted by $G + H$, is the graph with vertex set consisting of the disjoint union $V(G) \cup V(H)$ and edge set $E(G + H) = E(G) \cup E(H)\{uv | u \in V(G), u \in V(G)\}$. The graph S_k^+ is obtained by joining a vertex v to the union of a copy of K_2 and $k - 2$ copies of K_1. That is,

$$S_k^+ = K_1 + (K_2 \cup (k - 2)K_1).$$

A copy of S_k^+ is said to be *centered at* v. When $k = 2$, we have $S_k^+ = K_3$. Alternatively, we can also define S_k^+ to be the graph obtained from the star $K_{1,k}$ by adding a single edge between two of the pendant vertices. See Fig. 1.3 for an illustration of S_4^+.

Fig. 1.4 A graph and its complement

$$G \qquad \overline{G}$$

A subset of vertices in a graph G is called *independent* if its induced subgraph of G contains no edge. An independent set of vertices is also called an *empty* graph. The *independence number* of a graph G, denoted by $\alpha(G)$, is the maximum number of vertices in an independent set of G. A related concept for edges, a *matching* is a graph consisting of a vertex-disjoint collection of edges and their corresponding vertices.

The *complement* of a graph G, denoted by $\overline{G}$, is the graph on the same vertex set as G with an edge uv if and only if $uv \notin E(G)$. For example, the complement of a C_5 is again a C_5 while the complement of K_n is an empty graph on n vertices. See Fig. 1.4 for an example of an arbitrary graph G and its complement.

A graph G is *connected* if there exists a path, as a subgraph of G, between every pair of vertices in G. Otherwise, if there exists a pair of vertices with no path in G connecting them, then G is called *disconnected*. More generally, for a positive integer k, a graph is *k-connected* if the removal of any set of $k-1$ vertices from $V(G)$ leaves behind a connected graph. Equivalently and perhaps more appropriate for our discussions, a graph is k-connected if and only if between every pair of vertices in G, there exist at least k internally disjoint paths, proved by Menger [3], meaning that the paths are vertex-disjoint except at their ends. In particular, every cycle is 2-connected, the graph G in Fig. 1.3 is also 2-connected, but $\overline{G}$ is not 2-connected (although it is connected). A vertex in a 1-connected graph is called a *cut vertex* if its removal disconnects the graph. A maximal connected subgraph of a disconnected graph is called a *component* of the graph. A maximal 2-connected subgraph of a 1-connected graph is called a *block*.

Similarly, a graph G is called *k-edge-connected* if the removal of any set of $k-1$ edges leaves behind a connected graph. In particular, if a graph is 1-edge-connected, an edge whose removal disconnects the graph is called a *cut edge* or a *bridge*. For example, every cycle is 2-edge-connected.

Since this work concerns edge-colored graphs, we also define the *color degree* of a vertex $v \in V(G)$, denoted by $d^c(v)$, to be the number of different colors on edges incident to v. The *minimum color degree*, denoted by $\delta^c(G)$, (and *maximum color degree*, denoted by $\Delta^c(G)$), is then defined to be the minimum (and the maximum) of all the color degrees of vertices in $V(G)$.

Often we consider functions of graph properties like the number of available colors or the order of the graph. The behavior of these functions will sometimes be summarized using *little-o notation* as follows. For two functions f and g of a variable n, say $f(n) = o(g(n))$ if $\lim_{n \to \infty} \frac{f(n)}{g(n)} = 0$.

1.2 Ramsey Definitions

Given a positive integer k and a graph H, let $R_k(H)$ denote the k-color *Ramsey number* for finding a monochromatic copy of H, that is, the minimum integer N such that every k-coloring of the edges of K_n, where $n \geq N$, contains a monochromatic copy of H. When $k = 2$, we use the more common notation $R(H, H) = R_2(H)$. More generally, given k (possibly different) graphs $H_1, H_2, \ldots, H_k$, define the k-color Ramsey number for these graphs, denoted by $R(H_1, H_2, \ldots, H_k)$, to be the minimum integer N such that any k-coloring, using colors from $[k]$, of the edges of K_n, where $n \geq N$, contains a monochromatic copy of H_i in color i for some i with $1 \leq i \leq k$.

There is a vast body of literature on Ramsey Theory which we make no attempt to thoroughly survey in this work. We refer the interested reader to the outstanding dynamic survey by Radziszowski [4] that we frequently consult for an up-to-date account of known Ramsey results. Since many of the results in this text depend heavily upon Ramsey numbers, we will state the known results (found in [4]) as needed throughout the text.

We now present the main definition of this text, a restriction, in a sense, of Ramsey numbers to the class of colored complete graphs free of a specified rainbow-colored subgraph.

Definition 1.1 Given non-empty graphs G and H, define the *Gallai-Ramsey number* $gr_k(G : H)$ to be the minimum integer N such that for all $n \geq N$, every coloring of K_n, using at most k colors, contains either a rainbow colored copy of G or a monochromatic copy of H.

In the literature, a colored complete graph is called a *Gallai coloring* if it contains no rainbow triangle. The definition of the Gallai-Ramsey number is more flexible in that the rainbow graph can be anything, not just a triangle.

Note that the Gallai-Ramsey numbers are guaranteed to exist since the Ramsey number $R_k(H)$ is known to exist (see [5]) and since Gallai-Ramsey numbers consider only restricted colorings of K_n, we then have

$$gr_k(G : H) \leq R_k(H) < \infty.$$

Then if $k < |E(G)|$, then $gr_k(G : H) = R_k(H)$ since there can be no rainbow copy of G with only k colors. In general though, when $k \geq |E(G)|$, the relationship between $gr_k(G : H)$ and $R_k(H)$ is much more interesting.

As an easy example to get a feeling for the Gallai-Ramsey numbers we offer the following simple result.

Theorem 1.1 *For $k \geq 1$, we have*

$$gr_k(K_3 : P_3) = 3.$$

Proof For the lower bound, no matter how we color the single edge of K_2, there can be no rainbow triangle or monochromatic copy of P_3. This means that

$$gr_k(K_3 : P_3) > 2.$$

For the upper bound, we consider an arbitrary coloring of the triangle K_3 with any number of available colors. If three colors appear in this triangle, then it is a rainbow-colored triangle and we have thus produced one of the desired colored subgraphs. There must therefore be at most two colors appearing in the triangle. This means that two edges of the triangle must have the same color (while the third edge might or might not also share this color). This produces a monochromatic copy of P_3, the other desired colored subgraph, completing the proof.   □

As trivial as the above proof is, it demonstrates the overview of the (standard) proof strategy employed in most sharp Ramsey-type arguments. One must first show that the number is at least some value and then show that the number is at most some value. These two arguments are often very different in nature. For the lower bound, one must produce an example coloring, a coloring that does not contain the desired colored subgraphs but has as many vertices as possible. The sharp number will be one more than the number of vertices in this example. For the upper bound, one must argue that with one more vertex, regardless of the coloring, one of the desired colored subgraphs must appear.

In the study of Gallai-Ramsey numbers, the general logic for the upper bounds is to forbid the rainbow-colored subgraph and use the resulting structure to force the existence of a copy of the monochromatic subgraph. The proofs then naturally follow these steps:

1. Construct an example coloring of a complete graph, free of a rainbow copy of G or a monochromatic copy of H, on as many vertices as possible, to serve as a lower bound.
2. Describe the structure of all colored complete graphs which contain no rainbow copy of G in as much detail as possible.
3. Use the described structure to force, in sufficiently large colored complete graphs, the existence of a monochromatic copy of H.

As with Ramsey numbers looking for different monochromatic graphs in the different colors, define the *refined k-colored Gallai-Ramsey number*, denoted by $gr_k(G : H_1, H_2, \ldots, H_k)$, to be the minimum number of vertices N such that every k-coloring (using at most k colors) of the complete graph on $n \geq N$ vertices contains either a rainbow colored G, or a copy of H_i in color i, for some i with $1 \leq i \leq k$.

This refined notion is most often used, for the sake of applying induction on k, when proving more precise results. For example, when proving Theorem 3.19 about $gr_k(K_3 : K_4)$, we actually prove the more refined version in Theorem 3.20 about $gr_k(K_3 : K_4, K_4, \ldots, K_4, K_3, K_3, \ldots, K_3)$. The proof strategy examines what happens if we remove one vertex at a time with all one color on its incident edges to the remaining vertices, say a vertex v with all incident edges being red. In the remaining subgraph restricted to the red edges, it would suffice to find a red triangle, which would then (along with v) produce a monochromatic copy of K_4 in the original graph.

References

1. G. Chartrand, L. Lesniak, P. Zhang, *Graphs and Digraphs*, 5th edn. (CRC Press, Boca Raton, FL, 2011)
2. X. Li, C. Magnant, Z. Qin, *Properly Colored Connectivity of Graphs*, SpringerBriefs in Mathematics (Springer, Cham, 2018)
3. K. Menger, Zur allgemeinen Kurventheorie. Fund. Math. **10**, 96–115 (1927)
4. S.P. Radziszowski, Small Ramsey numbers. Electron. J. Combin. **1**; Dyn Surv **1**, 30 (1994). (electronic)
5. F.P. Ramsey, On a problem of formal logic. Proc. Lond. Math. Soc. **30**, 264–286 (1930)

Chapter 2
General Structure Under Forbidden Rainbow Subgraphs

The general structure of colored complete graphs containing no copy of a particular rainbow subgraph has been extremely useful in establishing sharp Ramsey-type results for finding monochromatic subgraphs. Several small graphs, like P_3 for example, immediately trivialize the problem. Indeed, if a colored complete graph contains no rainbow copy of P_3, then it must be colored entirely with one color. Adding in the third edge to make a triangle already makes the problem much more interesting. When a rainbow subgraph G is forbidden from a coloring, we say the coloring is *rainbow G-free*. For example, if a rainbow triangle is forbidden, we say the coloring is *rainbow triangle-free*.

2.1 Rainbow Triangles

A *Gallai coloring* of a complete graph G is an edge-coloring of G such that G does not contain a rainbow triangle as a subgraph. This definition is so named in honor of the following classical result, originally by Gallai from 1967. Here a partition of the vertices is called *trivial* if it partitions the vertex set into only one part. Thus, a nontrivial partition breaks the vertices of the graph K_n into t parts where $2 \leq t \leq n$.

This result is attributed to Gallai [1] but can also be traced to Cameron et al. [2]. Gallai's original statement concerned transitive orientations of graphs. Gyárfás and Simonyi published in [3] a translation of Gallai's result to the context of edge-colored complete graphs.

Theorem 2.1 ([1–3]) *Every Gallai coloring of a complete graph K_n has a nontrivial partition (that is, with at least two parts) of the vertices such that in between the parts, there are a total of at most two colors on the edges, and in between each pair of parts, there is only one color on the edges.*

© The Author(s), under exclusive license to Springer Nature Switzerland AG 2020
C. Magnant and P. Salehi Nowbandegani, *Topics in Gallai-Ramsey Theory*,
SpringerBriefs in Mathematics, https://doi.org/10.1007/978-3-030-48897-0_2

Fig. 2.1 An example of a
Gallai partition

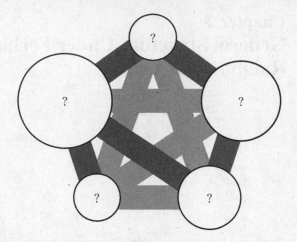

Such a partition is called a *Gallai partition* and we define the *reduced graph* of a
Gallai partition to be the subgraph of the original Gallai coloring induced on a single
vertex from each part of the Gallai partition. By Theorem 2.1, the reduced graph is a
2-colored complete graph. More precisely for a given Gallai partition $P : A_1, \ldots, A_t$,
we define $R = R(G)$ to be the 2-colored complete graph on t vertices, one vertex
with for each part of the partition, and the color of the edge between two vertices be
the same as the color between their corresponding parts in G.

Note that if the Gallai partition was trivial in Theorem 2.1, the conclusion of the
theorem would be vacuous. At the opposite extreme, if the Gallai partition breaks
the graph into n parts each of order 1, then the coloring uses at most 2 colors and the
fact that this coloring contains no rainbow triangle becomes obvious. See Fig. 2.1 for
an example of a Gallai partition. Note that this result leaves several questions only
partially answered:

- There is no indication how many parts there are, although the number is between
  2 and n, inclusive;
- There is no indication how big the parts are, although they are disjoint and the
  orders sum to n;
- There is no indication what happens with the edges inside each part, although they
  do form a Gallai coloring since there is no rainbow triangle in the whole graph,
  so certainly no rainbow triangle within any part. This means that there is another
  Gallai partition within each part, but these partitions may use different pairs of
  colors.

It is easy to see that Theorem 2.1 is equivalent to the following Theorem 2.2. We
will prove Theorem 2.2 by following the outline of the proof presented by Gyárfás
and Simonyi in [3]. For this statement, by "substituting" a complete graph H of order
m into a vertex $v \in V(G)$, we mean the graph obtained from G by making m copies
of v (copying all of its edges as well) and inserting the edges of H between the copies
of v. See Fig. 2.2 for an illustration of this.

Fig. 2.2 An example of
substituting

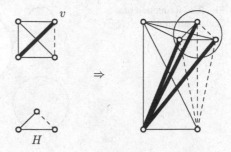

Theorem 2.2 ([1–3]) *Every Gallai coloring of a complete graph can be obtained by substituting complete graphs with Gallai colorings into the vertices of a 2-colored complete graph.*

We follow the proof presented by Gyárfás and Simonyi in [3]. First a helpful lemma.

Lemma 2.1 ([3]) *Every Gallai coloring of a complete graph using at least 3 colors has a color that spans a disconnected graph.*

For example, if red is a color that is disconnected, this means that the subgraph containing all the vertices of the graph and precisely the red edges of the graph is disconnected. In particular, if a vertex has no red edges, then it is, by itself, a component of the subgraph.

With this lemma in hand, we first prove Theorem 2.2 before coming back to prove Lemma 2.1.

Proof *(of Theorem 2.2)* Let G be a Gallai coloring of a complete graph K_n. We prove this result by induction on n. The base of this induction is when $n \leq 3$, in which case the result is trivial. In particular, if $n = 3$, since there is no rainbow triangle, at least two edges of this triangle must share a color, meaning that G itself is a 2-colored triangle.

Suppose $n \geq 4$. If G uses at most 2 colors, the result is immediate since G itself is again a 2-colored complete graph. Thus, suppose G uses at least 3 colors and, for each color i, let G_i be the spanning subgraph of G containing precisely the edges of color i.

By Lemma 2.1, there is a color i such that G_i is disconnected. Without loss of generality, suppose $i = 1$. Let $H_1, H_2, \ldots, H_t$ be the components of G_1, say with $1 \leq |H_1| \leq |H_2| \leq \cdots \leq |H_t|$. Since color 1 appears in G, we know that $|H_t| \geq 2$. Note that the subgraph of G induced on each set H_i is itself a Gallai coloring since there can be no rainbow triangle anywhere within G.

In order to avoid a rainbow triangle, all edges between H_j and H_t must have a single color for each j with $1 \leq j < t$. The set H_t may then be reduced down to a single vertex, merging all parallel edges into a single edge of the same color, to create a new graph G'. See Fig. 2.3 for an example of this reduction to G'. Note that

Fig. 2.3 Reduction of a
Gallai partition

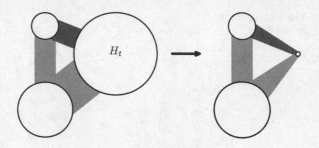

$|G'| < |G|$ and G' is also a Gallai coloring of a complete graph. By induction on n, we see that G' can be constructed by substituting Gallai colored complete graphs into the vertices of a 2-colored complete graph. Since G is created from G' by the same substitution, we see that G can also be constructed by substituting Gallai colored complete graphs into the vertices of a 2-colored complete graph, to complete the proof.   □

Proof (of Lemma 2.1) Let G be a vertex-minimal counterexample to Lemma 2.1, say with k colors where $k \geq 3$. Each vertex of G has at least one incident edge in all k colors since any color missing from a vertex must necessarily be disconnected. Let $x \in V(G)$ and let $H = G \setminus \{x\}$. Then H cannot be 2-colored because then any other color would have been disconnected in G. For each color i, let H_i denote the spanning subgraph of H containing precisely the edges of color i. By the minimality assumption, H_i must be disconnected for some color i. Without loss of generality, suppose $i = 1$.

Let $C_1, C_2, \ldots, C_t$ be the components of H_1. As in the proof of Theorem 2.2, we get the following fact.

Fact 2.3 *All edges between any pair of components C_i and C_j must have a single color (different from color 1).*

We claim that color 1 is also disconnected in G, which would contradict our assumptions and complete the proof.

Claim 2.4 *The spanning subgraph of G containing precisely the edges in color 1 is disconnected.*

Proof For a contradiction, suppose there is at least one edge in color 1 from x to some vertex $y_i \in V(C_i)$ for each i. Since x must also have an incident edge of every other color, let xu and xv be edges of colors 2 and 3, respectively.

First suppose u and v are both in a single component C_i, say with $i = 1$. Then uy_2 must have color 2 and vy_2 must have color 3 to avoid creating a rainbow triangle on xuy_2 and xvy_2, respectively. See Fig. 2.4 for a picture of this structure. This contradicts Fact 2.3.

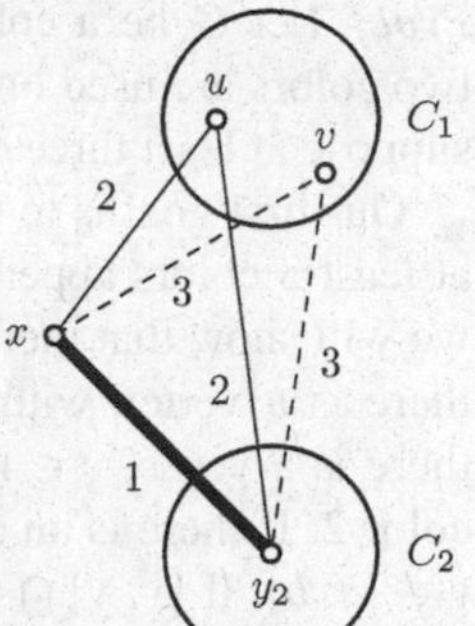

Fig. 2.4 Structure when u and v are both in C_1

Fig. 2.5 Structure when u and v are in different components

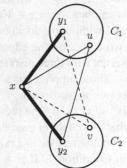

Finally suppose u and v are in different components, say with $u \in C_1$ and $v \in C_2$. Then uy_2 must have color 2 while vy_1 must have color 3 to avoid creating a rainbow triangle on xuy_2 and xvy_1, respectively. See Fig. 2.5 for a picture of this structure. This again contradicts Fact 2.3, completing the proof of Claim 2.4. $\qquad\square$

As noted above, Claim 2.4 completes the proof of Lemma 2.1. $\qquad\square$

2.2 Rainbow Paths

If a rainbow path is forbidden, Thomason and Wagner [4] provided some beautiful structural results. For P_4, we include the proof adapted from [4] by, but we do not include the proof for P_5 since it is longer and technical. Note that every coloring of a complete graph using at least 2 colors contains a rainbow P_3 so forbidding a rainbow P_t only makes sense for $t \geq 4$.

Theorem 2.5 ([4]) *Let $n \geq 4$ and let G be a coloring of K_n containing no rainbow copy of P_4. Then either*

1. *at most two colors appear on the edges of G, or*
2. *$n = 4$ and three colors are used, each forming a perfect matching.*

Proof Let G be a coloring of K_n containing no rainbow copy of P_4. If at most two colors are used on the edges, this satisfies the first item so, for a contradiction, suppose at least three colors appear.

Our first goal is to show that there is a vertex with color degree at least 3. With at least 3 colors appearing, there must be a vertex with color degree at least 2 but we will show that there cannot be a vertex of color degree exactly 2, implying that there is a vertex with color degree at least 3. Suppose, for a contradiction, that there is a vertex $v \in V(G)$ with color degree 2, say with vx in color 1 and vy in color 2. If there is an edge $e = ab \in E(G)$ with a third color, then by assumption, $v \notin \{a, b\}$. If $\{x, y\} \cap \{a, b\} = \emptyset$, then either $xvab$ or $yvab$ forms a rainbow P_4. We may therefore assume $\{x, y\} \cap \{a, b\} \neq \emptyset$. Then, to avoid a rainbow P_4, we must have $\{x, y\} = \{a, b\}$, meaning that vxy forms a rainbow triangle. Since $n \geq 4$, there is another vertex $z \in V(G) \setminus \{v, x, y\}$. The edge vz cannot be colored with color 1 or 2 without producing a rainbow P_4, a contradiction. Thus, either there are at most two colors on the edges of G, meaning that G satisfies Condition 1, or there is a vertex with three distinct colors.

Thus, suppose v is a vertex with color degree 3, say with edges of three different colors to vertices w, x, y. Let H be the subgraph of G induced on the vertices $\{v, w, x, y\}$. By the previous argument, no vertex in H has color degree 2. Suppose that two edges of a single color share a vertex, say w. Then w must have color degree 1 but then $xvyw$ is a rainbow P_4. This means that H must have the form described in Condition 2. If $n \geq 5$, then there is a vertex $z \in G \setminus \{v, w, x, y\}$ but the edge vz cannot be colored without making a rainbow P_4. Thus, $n = 4$ and Condition 2 holds. $\qquad\square$

For this next statement, we define some additional notation. Within a given colored graph G, let E_i denote the set of edges in G with color i.

Theorem 2.6 ([4]) *Let $n \geq 5$ and let G be a coloring of K_n containing no rainbow copy of P_5. Then, up to relabeling the colors, one of the following must hold:*

1. *at most three colors are used,*
2. *there is a universal color, say color 1, such that no vertex has incident edges in two different colors other than color 1,*
3. *$G \setminus \{v\}$ is monochromatic for some $v \in V(G)$,*
4. *there are three vertices a, b, c such that $E_2 = \{ab\}$, $E_3 = \{ac\}$, E_4 consists of bc and perhaps some edges incident to a, and all remaining edges have color 1,*
5. *there are four vertices a, b, c, d such that $\{ab\} \subseteq E_2 \subseteq \{ab, cd\}$, $E_3 = \{ac, bd\}$, $E_4 = \{ad, bc\}$ and all remaining edges have color 1, or*
6. *$n = 5$, $V(G) = \{a, b, c, d, e\}$, $E_1 = \{ad, ae, bc\}$, $E_2 = \{bd, be, ac\}$, $E_3 = \{cd, ce, ab\}$ and $E_4 = \{de\}$.*

2.3 Rainbow Triangle with a Pendant Edge

Recall that S_3^+ is a claw $K_{1,3}$ with the addition of an edge between two of the leaves. Equivalently, this can be seen as a triangle with the addition of a pendant edge. The goal of this section is to provide a structural result similar to Theorem 2.1 for rainbow S_3^+-free colorings. Unfortunately, in light of the following two examples, the partition cannot be as strong as the one provided by Theorem 2.1.

First consider the example in Fig. 2.6A. Let $G = A_1 \cup A_2 \cup A_3 \cup A_4$ where each A_i for $i = 1, 2, 3$ is a complete graph on color i and A_4 is any rainbow triangle-free coloring of a complete graph (using any number of colors). The edges between A_i and A_4 have color i for $i = 1, 2, 3$. Also, the edges between A_i and A_j have colors i and j and we require that there exists a spanning tree in each of these colors within the bipartite graph induced on $A_i \cup A_j$. This graph contains no rainbow S_3^+ but also contains no partition of the vertices in which there are a total of at most two colors on the edges between the parts.

Furthermore, consider the example in Fig. 2.6B. To complete the construction, let all edges not pictured have color 1. This graph contains no rainbow S_3^+, uses $2k' + 1$ colors on $4k' - 1$ vertices and has no partition such that there is only one color between each pair of parts (aside from the partition in which each vertex is its own part). Note that we may add vertices without adding more colors by extending one of the pendants in the corresponding color.

Fortunately, as seen in the next theorem, these two problems cannot occur at the same time.

Theorem 2.7 ([5]) *In any rainbow S_3^+-free coloring G of a complete graph, one of the following holds:*

1. *$V(G)$ can be partitioned such that there are at most 2 colors on the edges in between the parts; or*
2. *There are three (different colored) monochromatic spanning trees, and moreover, there exists a partition of $V(G)$ with exactly 3 colors on edges between parts and between each pair of parts, the edges have only one color.*

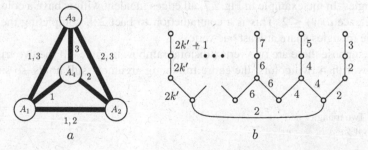

Fig. 2.6 Constructions

Note in item 1 above, both of the colors may appear on edges between a single pair of parts of the partition. This means that any partition with at most 2 colors on the edges between parts (as in Case 1 of Theorem 2.7) implies the existence of a bipartition, also with at most two colors on the edges between parts. Included below is the proof of Theorem 2.7 by Fujita and Magnant [5] expanded to include more detail.

Proof Suppose there exists a rainbow S_3^+-free coloring of K_n such that there is no partition of the vertices as stated. We first provide some facts and claims that will eventually lead to a contradiction.

Fact 2.8 *For any rainbow triangle T in G using colors in $C = \{c_1, c_2, c_3\}$, all edges from the vertices of T to $G \setminus T$ must have a color in C.*

This fact follows from the observation that, if there was an edge e in a 4th color coming out of T, then $T \cup e$ forms a rainbow copy of S_3^+.

Fact 2.9 *There exists no vertex v with color degree $d^c(v) \leq 2$, that is, $\delta^c(G) \geq 3$.*

The proof of this fact involves a construction of the desired partition. Consider the partition of the vertices of $V(G)$ into $\{v\}$ and $G \setminus \{v\}$. Since v has color degree $d \leq 2$, there are d colors between the two parts of this partition, providing the desired partition.

Fact 2.10 *There exists a rainbow triangle.*

If there exists no rainbow triangle, then by Theorem 2.1, the graph G has a Gallai partition, even more restrictive than the desired partition. Next, we show two claims about even more detailed structure.

Claim 2.11 *There exist no two rainbow triangles sharing at most two colors.*

Proof First suppose there exist two rainbow triangles sharing at most two colors and sharing at least one vertex. Let w be a vertex shared between the triangles (regardless of whether these triangles share one or two vertices). See Fig. 2.7. By Fact 2.8, since w is in both triangles, w must only have incident edges in the colors shared between the triangles. In the example in Fig. 2.7, all edges incident with w have a color from $\{1, 2\}$. Hence $d(w) \leq 2$. This is a contradiction to Fact 2.9, completing the proof when the triangles share at least one vertex.

Next suppose there are two vertex-disjoint rainbow triangles. If the two triangles share fewer than two colors, the entire following argument still holds, so suppose

Fig. 2.7 Two triangles and the vertex w

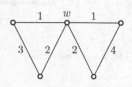

there exist two vertex-disjoint rainbow triangles $T_1 = u_1u_2u_3$ and $T_2 = v_1v_2v_3$ with colors 1, 2, 3 and 2, 3, 4, respectively. By Fact 2.8, all edges between T_1 and T_2 must have color 2 or 3 (call these the *shared colors*).

Certainly if $G = T_1 \cup T_2$, then the partition into T_1 and T_2 is the desired partition so suppose there exists a vertex $v \in G \setminus (T_1 \cup T_2)$. See Fig. 2.8.

We now show that v must have only edges of the shared colors to at least one of the triangles T_i. For a contradiction, suppose v has an edge of a non-shared color to each of the triangles. By Fact 2.8, this means that v has an edge of color 1 to T_1 and an edge of color 4 to T_2. Without loss of generality, suppose v is adjacent to $u_1 \in T_1$ in color 1 and $v_2 \in T_2$ in color 4. This means that v is on a rainbow triangle since the edge u_1v_2 must have a shared color. Hence, vu_1v_2 and T_1 are two rainbow triangles that share a vertex and share only two colors which contradicts Claim 2.11. This means that v must have only shared colors to at least one of T_1 or T_2.

Let H_1 be the set of vertices v such that v has at least one edge of color 1 to T_1. With this definition and the previous argument, it follows immediately that T_2 has only edges of shared colors to H_1. Next define a sequence of sets H_j. For convenience, let $H_0 = T_1$ and given sets $\{H_0, H_1, \ldots, H_{j-1}\}$, define H_j to be the set of vertices $v \in G \setminus (H_0 \cup H_1 \cup \cdots \cup H_{j-1})$ such that v has at least one edge of color 1 to H_{j-1}. In other words, the set H_j is the set of vertices at distance exactly j from T_1 within the subgraph induced on color 1.

Let i be the smallest index of a set H_i such that there exists a vertex $v \in H_i$, a vertex $x \in H_{i+1}$ with $c(vx) = 1$, and a vertex $y \in G \setminus (H_0 \cup \cdots \cup H_i)$ with $c(vy) = 2$ or 3 (suppose 2) and $c(xy) = c > 3$. See Fig. 2.9 for a picture of this structure. Since vxy forms a rainbow triangle with colors 1, 2, c and $c \neq 3$, by Claim 2.11, this implies $i \geq 1$.

We first show that such an index i must exist, otherwise we show that there exists a partition as desired. Suppose there exists no vertex $x' \in H_{i'+1}$ with an edge of a

Fig. 2.8 Two triangles and the vertex v

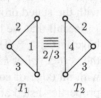

Fig. 2.9 Structure of v, x and y

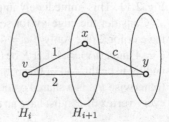

Fig. 2.10 Structure of v', x' and y'

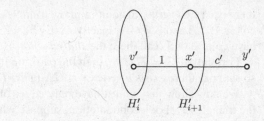

Fig. 2.11 Structure of v', v, x and y

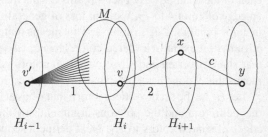

color c' with $c' > 3$ to a vertex $y' \in G \setminus (H_0 \cup \cdots \cup H_{i'+1})$ for any index i' (note that here y' is not allowed to be in the same set $H_{i'+1}$ as x' whereas the minimum such index i may allow $y \in H_{i+1}$). By definition, we have $T_2 \cap H_j = \emptyset$ for all j so consider the partition of G into $A = H_0 \cup H_1 \cup \ldots$ and $B = G \setminus (H_0 \cup H_1 \cup \ldots)$. Note that $|B| \geq 3$ since $T_2 \subseteq B$. By the definition of H_j and the assumption that no vertex x' exists, we see that all edges between A and B must have colors 2 or 3, which means this is the desired partition. Hence, such a vertex x' must exist, see Fig. 2.10. Suppose x' is chosen from the set $H_{i'+1}$ for the smallest value of i' and let y' be the opposite end of the edge of color $c' > 3$ from x'. Such an edge must exist by the choice of x'. If we let v' be a vertex in $H_{i'}$ with $c(v'x') = 1$, then the edge $v'y'$ must not have color 1 since $y' \notin H_{i'+1}$. Furthermore, by the minimality of the index i', it must not have a color greater than 3. Hence, $v'x'y'$ forms a rainbow triangle with colors 1, c' and some shared color. This proves that there must exist an index with the stated properties, so we choose i to be the smallest such index.

Recall that we have observed that v is in a rainbow triangle and by Fact 2.9, we have $\delta^c(G) \geq 3$. These facts imply that there must exist an edge of color c containing the vertex v. Let M be the component of G_c containing v. Let v' be a vertex in H_{i-1} with an edge of color 1 to v. Since i was chosen to be the smallest index with the properties listed above, we find that all edges from v' to M must have color 1. See Fig. 2.11. This immediately implies that $T_2 \cap M = \emptyset$.

Consider the edge uv for some $u \in T_2$. By Fact 2.8, we see that this edge must have color 2 or c (only if $c = 4$). We would like to show that $c(uv) = 2$ so suppose $c = 4$ and $c(uv) = 4$. Let P be the shortest path (containing v') in color 1 from v to T_1 and let w be the end of this path with $w \in T_1$. By Claim 2.11, we see that $c(uv') = 4$ (otherwise uv' is a shared color and uvv' would form a rainbow triangle). Similarly, every vertex of P must have an edge of color 4 to u. In particular, $c(uw) = 4$ but this, along with T_1 forms a rainbow copy of S_3^+, for a contradiction. Hence, we must have $c(uv) = 2$. If u is contained in a rainbow triangle with two vertices of M, then by Claim 2.11, this must have colors 2, 3, 4. Since v' has all edges of color 1 to M, this creates a rainbow S_3^+, a contradiction. Hence, since $u \notin M$, u must have all edges of color 2 to M.

We have forced every vertex in M to have degree at least 3 (using colors 1, 2, c). By Fact 2.8, we see that every rainbow triangle which includes a vertex of M must use colors 1, 2, c. We now claim that M and $G \setminus M$ form the desired partition with colors 1 and 2 between the parts. For a contradiction, suppose there exists an edge of color d from a vertex $u_1 \in M$ to $u_2 \in G \setminus M$ where $d \geq 3$. By the definition of M, we see that $d \neq c$. Since there is no rainbow triangle including a vertex of M which has color d, this means that all edges from u_2 to M must have color d. In particular, the edge from u_2 to v must have color d, but this forms a rainbow S_3^+ on the vertices u_2vxy. This contradiction completes the proof of the claim.

If there exist two rainbow triangles which share only one color, the same argument can be applied for a contradiction.   □

With these claims in hand, we are now able to proceed with the proof of Theorem 2.7.

By Fact 2.10, there must exist a rainbow triangle in G. Let T be such a rainbow triangle, say in colors 1, 2 and 3. If the entire graph is colored with three colors and each of the graphs induced on single colors contains a spanning tree of G, then each vertex can be its own part of a partition which satisfies our condition in the second part of the statement. Furthermore, if the entire graph is colored with three colors and at most two of the graphs induced on single colors contain a spanning tree of G, then we can define a partition based on the components of the third color to get a partition in which there are only two colors between the parts.

Then we may assume that there exists an edge uv of color 4. By Fact 2.8, we have $u, v \in G \setminus T$. By Claim 2.11, the edge uv cannot be used in a rainbow triangle.

First suppose there exist at most two colors which contain spanning trees. Clearly these two colors must be in the set $\{1, 2, 3\}$ since all edges incident to vertices in T come from $\{1, 2, 3\}$, so suppose color 1 does not contain a spanning tree. Let $C_1, C_2, \ldots$ be the components of G_1. Without loss of generality, suppose the edge of T in color 1 is in C_1.

Consider the bipartition of G into C_1 and $G \setminus C_1$. By definition, no edges between the parts have color 1. If all the edges between the parts had colors 2 or 3, this would be our desired partition, so suppose there exists an edge of some color greater than 3 (suppose color 4) between C_1 and a vertex $w \in G \setminus C_1$. By Claim 2.11, and Fact 2.8, the vertex w must not be on a rainbow triangle. Since there exists a tree in color 1 which spans C_1, all edges from w to C_1 must then have color 4. In particular, the edges from w to at least one vertex of T have color 4, but this creates a rainbow copy of S_3^+, for a contradiction.

Hence, we may suppose there exist three colors containing spanning trees and these colors must be 1, 2, 3. We will now prove that there must exist a partition such that there are at most 3 colors on edges between the parts and exactly one color on the edges between each pair of parts. Note that $G_{\geq 4} = \cup_{i \geq 4} G_i$ is disconnected since each vertex of T is a component of $G_{\geq 4}$.

Let C be the component of $G_{\geq 4}$ containing the edge uv (recall uv has color 4). Consider a vertex $w \in G \setminus C$. The vertex w has only edges of colors 1, 2 or 3 to C so suppose there is an edge of color i to C for some $i \leq 3$. By Claim 2.11, there is not a rainbow triangle using w and an edge of $G_{\geq 4}$. In order to avoid such a rainbow triangle, all edges from w to C must have color i.

We now build our desired partition. Consider the partition of the graph into the components $H_1, H_2, \ldots, H_t$ of $G_{\geq 4}$. Certainly this is a nontrivial partition since the vertices of T must have only edges of colors 1, 2 and 3, so these vertices each form their own parts of this partition. By definition, between any two parts, there are only edges of colors 1, 2 or 3.

If a component H_i contains at least one edge, then there is a spanning tree of H_i in $G_{\geq 4}$. By Claim 2.11, there does not exist a rainbow triangle using an edge of $G_{\geq 4}$. To avoid such a rainbow triangle, every edge to H_i from H_j for $i \neq j$, must have a single color. Hence, under the assumed restrictions, there exists a partition of G such that between parts, there are at most 3 colors and between any pair of parts, there is exactly one color. This completes the proof of Theorem 2.7.   □

Theorem 2.7 implies that, in any rainbow S_3^+-free coloring, there is a partition with at most 3 colors on the edges between the parts. As an immediate corollary, we get the following extension of Lemma 2.1.

Corollary 2.1 *Every coloring of a complete graph containing no rainbow S_3^+ using at least 4 colors has a color that spans a disconnected graph.*

2.4 Other Rainbow Graphs

The first result of this section provides a generalization of Theorem 2.7 to a larger star with an additional edge. Recall that S_k^+ is constructed from a star $K_{1,k}$ by adding a single edge between two leaves of the star.

Theorem 2.12 ([5]) *For $k \geq 4$, in any rainbow S_k^+-free coloring G of a complete graph, there exists a partition of $V(G)$ such that between the parts, there are at most k colors. Furthermore, this result is sharp since there exists such a coloring in which k colors appear on edges between the parts of every partition of the vertices.*

Proof Let G be a rainbow S_k^+-free coloring of a complete graph and suppose G has no such partition of the vertices. Certainly we may assume there exists a rainbow triangle in G since, otherwise, Theorem 2.1 would imply a partition in which only 2 colors appear on edges between the parts and hence we have a stronger partition. If there exists a vertex v with at most k incident edges, then v and $G \setminus \{v\}$ form the desired partition. Hence, we may assume that every vertex has at least $k + 1$ different colors on incident edges. Consider a rainbow triangle T in G and let $v \in T$. Since v has at least $k + 1$ different colors on incident edges, there exists a rainbow S_k^+ centered at the vertex v. This contradicts the choice of G.

For sharpness, we construct a k coloring of K_n such that for every partition of $V(G)$, there are k colors between the parts of the partition. Let n be large enough such that there exist k edge-disjoint hamiltonian cycles in K_n. Any number at least $2k + 1$ (see [6]) would suffice. Color each hamiltonian cycle with a different color and color the remaining edges arbitrarily with the same k colors. This coloring uses a total of k colors on the edges of G so there exists no S_k^+ but certainly any partition of $V(G)$ will have all k colors appearing on edges between parts.   □

Next a strong result about graphs with no rainbow $K_{1,3}$. For this first statement, we need a definition. For $n \geq 1$, let $G^1(n)$ be a 3-edge-coloring of K_n that satisfies the following conditions: The vertices of K_n are partitioned into three pairwise disjoint sets V_1, V_2 and V_3 such that for $1 \leq i \leq 3$ (with indices modulo 3), all edges between V_i and V_{i+1} have color i, and all edges connecting pairs of vertices within V_{i+1} have color i or $i + 1$. Note that one of V_1, V_2 and V_3 is allowed to be empty, but at least two of them are non-empty. (Otherwise at most only two colors can appear.)

Theorem 2.13 ([7]) *For positive integers m and n, if G is an m-edge-coloring of K_n without rainbow $K_{1,3}$, then after renumbering the colors, one of the following holds:*

1. *$m \leq 2$ or $n \leq 3$,*
2. *$m = 3$ and $G \simeq G^1(n)$,*
3. *$m \geq 4$ and there is a universal color, say color 1, such that no vertex has incident edges in two different colors other than color 1.*

Essentially, this means that if G has no rainbow $K_{1,3}$, then either there is a universal color or the graph has a very prescribed form.

For our next result, we need two more definitions of specific graph structures.

For $n \geq 4$, let $G^2(n)$ be a 4-edge-coloring of K_n in which there is exactly one edge, say xy, having color 2. Every edge from x to all other vertices except y has color 3, and every edge from y to all other vertices except x have color 4. All edges not incident to vertices x, y have color 1. See Fig. 2.12. This graph contains no rainbow P_4^+ (a path on 4 vertices with the addition of a pendant vertex adjacent to an interior vertex of the path) but contains a rainbow $K_{1,3}$ and (if $n \geq 5$) a rainbow P_5.

For $n \geq 4$, let $G^3(n)$ be an arbitrary 4-edge-coloring of K_n in which there exists a rainbow K_3, say having colors 1, 2 and 3. Let every edge incident to at most one vertex in the rainbow K_3 have color 1. See Fig. 2.13. This graph contains no rainbow P_4^+ and no rainbow P_5, but contains a rainbow $K_{1,3}$.

Theorem 2.14 ([7, 8]) *For positive integers m and n, if G is an m-edge-coloring of K_n without a rainbow copy of P_4^+, then after renumbering the colors, one of the following holds:*

Fig. 2.12 $G^2(n)$

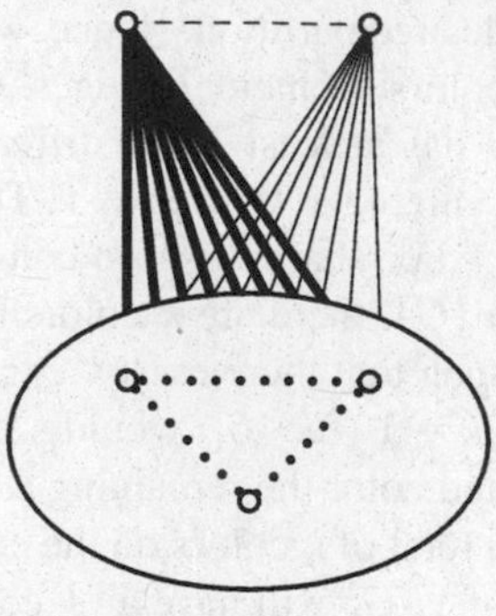

Fig. 2.13 $G^3(n)$

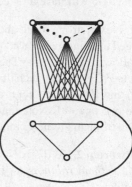

1. $m \leq 3$ or $n \leq 4$.
2. $m = 4$ and $G \in \{G^2(n), G^3(n)\}$.
3. $m \geq 4$ and G contains no rainbow $K_{1,3}$. In particular, there is a universal color, say color 1, such that no vertex has incident edges in two different colors other than color 1.

This again means that if G has no rainbow P_4^+, then either there is a universal color or the graph has a very prescribed form. Bass et al. [7] observed that, although P_4^+ contains $K_{1,3}$ and so forbidding a rainbow $K_{1,3}$ implies forbidding a rainbow P_4^+, the reverse implication also holds with only a finite number of counterexamples for each value of n.

References

1. T. Gallai, Transitiv orientierbare Graphen. Acta Math. Acad. Sci. Hungar **18**, 25–66 (1967)
2. K. Cameron, J. Edmonds, Lambda composition. J. Graph Theory **26**(1), 9–16 (1997)
3. A. Gyárfás, G. Simonyi, Edge colorings of complete graphs without tricolored triangles. J. Graph Theory **46**(3), 211–216 (2004)
4. A. Thomason, P. Wagner, Complete graphs with no rainbow path. J. Graph Theory **54**(3), 261–266 (2007)

5. S. Fujita, C. Magnant, Extensions of Gallai-Ramsey results. J. Graph Theory **70**(4), 404–426 (2012)
6. G. Chartrand, L. Lesniak, P. Zhang, *Graphs and Digraphs*, 5th edn. (CRC Press, Boca Raton, FL, 2011)
7. R. Bass, C. Magnant, K. Ozeki, B. Pyron, Characterizations of edge-colorings of complete graphs that forbid certain rainbow structures. *Manuscript*
8. A. Gyárfás, J. Lehel, R.H. Schelp, Zs. Tuza, Ramsey numbers for local colorings. *Graphs Combin.*, **3**(3):267–277 (1987)

Chapter 3
Gallai-Ramsey Results for Rainbow Triangles

3.1 General Bounds and Tools

Recall that a Gallai coloring of K_n is a rainbow triangle-free edge-coloring of K_n. For a connected bipartite graph H, define $s(H)$ to be the order of the smallest part in the unique bipartition of H and let $\ell(H)$ be the order of the largest part.

The first result of this section, by Wu et al., provides the lower bound for all Gallai-Ramsey numbers when the desired monochromatic graph is bipartite.

Proposition 3.1 ([1]) *For any connected bipartite graph H, and for any integer k with $k \geq 2$, we have*

$$gr_k(K_3 : H) \geq R(H, H) + (k - 2)(s(H) - 1).$$

Proof This result is proven by construction. Let G_2 be a sharpness example for the 2-color Ramsey number on $R(H, H) - 1$ vertices. For any $k > 2$, we assume that we have constructed a $(k - 1)$-coloring G_{k-1} of the complete graph on $R(H, H) + (k - 3)(s(H) - 1) - 1$ vertices with no rainbow triangle and no monochromatic H. Based on G_{k-1}, we construct G_k by adding a $s(H) - 1$ vertices with all incident edges in a new color (color k). The resulting graph G_k is a k-coloring of K_n with $n = R(H, H) + (k - 2)(s(H) - 1) - 1$ which contains no rainbow triangle and no monochromatic H. See Fig. 3.1. $\qquad\square$

For non-bipartite graphs, the picture is less clear. The following definition provides the base of the exponential lower bound for the Gallai-Ramsey numbers of non-bipartite graphs. Given a graph H, call a graph H' a *merge* of H if H' can be obtained from H by identifying sets of non-adjacent vertices (and removing any resulting repeated edges). Let $\mathcal{H}$ be the set of all possible merges of H. For the sake of the following definition, let $R_2(\mathcal{H})$ be the minimum integer n such that every 2-coloring of K_n contains a monochromatic copy of some graph in the set $\mathcal{H}$. Since

© The Author(s), under exclusive license to Springer Nature Switzerland AG 2020
C. Magnant and P. Salehi Nowbandegani, *Topics in Gallai-Ramsey Theory*,
SpringerBriefs in Mathematics, https://doi.org/10.1007/978-3-030-48897-0_3

Fig. 3.1 Construction of G_k

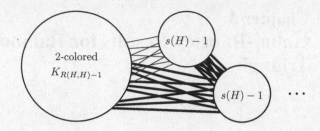

this quantity is bounded above by the Ramsey number $R(H, H)$, its existence is obvious. Now the main definition of this work.

Definition 3.1 ([2]) If $\mathscr{H}$ is the set of all merges of a given graph H, define the function $m(H)$ to be

$$m(H) = R_2(\mathscr{H}).$$

For example, if $H = K_n$, then the only merge of H is H itself so $m(K_n) = R_2(\mathscr{H}) = R(K_n, K_n)$. As a slightly less trivial example, consider the complete graph minus one edge $H = K_n - e$, say with $e = uv$. Then the only nontrivial merge of $K_n - e$ is K_{n-1} so $\mathscr{H} = \{K_{n-1}, K_n - e\}$. Since $K_{n-1} \subseteq K_n - e$, it is clear that

$$R_2(\{K_{n-1}, K_n - e\}) = R(K_{n-1}, K_{n-1})$$

so $m(K_n - e) = R(K_{n-1}, K_{n-1})$.

Using $m(H)$, we can obtain a general lower bound on all Gallai-Ramsey numbers for non-bipartite graphs H. Given colored complete graphs G_1 and G_2, a G_2-blow-up of G_1 is a new graph created by making $|G_2|$ copies of each vertex in G_1, also copying the incident colored edges, and inserting a copy of G_2 on the set of copies of each vertex. For example, if G_1 is a single red edge and G_2 is any colored graph, then a G_2-blow-up of G_1 would be two copies of G_2 with all red edges in between these copies. The *chromatic number* of a graph G, denoted by $\chi(G)$, is the smallest number of colors needed to color the vertices of G so that no two adjacent vertices share the same color.

Proposition 3.2 ([2]) *For a connected non-bipartite graph H and an integer $k \geq 2$, we have that $gr_k(K_3 : H)$ is at least*

$$\begin{cases} (R(H, H) - 1) \cdot (m(H) - 1)^{(k-2)/2} + 1 & \text{if } k \text{ is even}, \\ (\chi(H) - 1) \cdot (R(H, H) - 1) \cdot (m(H) - 1)^{(k-3)/2} + 1 & \text{if } k \text{ is odd}. \end{cases}$$

We include the proof by Magnant [2].

Proof This result is proven by an inductive construction. For the base of the induction, let G_2 be a 2-colored complete graph on n_2 vertices, where $n_2 = R(H, H) - 1$, containing no monochromatic copy of H. First note that, in this coloring, there is

Fig. 3.2 An example of this construction

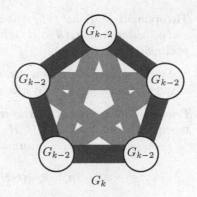

no rainbow triangle since only two colors are used. Such a coloring exists by the definition of the Ramsey number. Now suppose there is a $2i$-coloring G_{2i} of $K_{n_{2i}}$ on

$$n_{2i} = (R(H, H) - 1) \cdot (m(H) - 1)^{(2i-2)/2}$$

vertices containing no rainbow triangle and no monochromatic copy of H where $2i < k$.

First suppose $2i + 2 \leq k$. Let H_1 and H_2 be two reductions of H such that $m(H) = R(H_1, H_2)$ and let D be a 2-coloring of $K_{m(H)-1}$, using colors $i + 1$ and $i + 2$, containing no monochromatic copy of H_1 in color $i + 1$ and no monochromatic copy of H_2 in color $i + 2$. Now create a G_{2i}-blow-up of D. See Fig. 3.2 for an example of this construction. Since H_1 and H_2 are reductions of H, this blow-up of D contains no monochromatic copy of H. This means that the resulting graph, G_{2i+2} is a $(2i + 2)$-coloring of $K_{n_{2i+2}}$ on

$$n_{2i+2} = (R(H, H) - 1) \cdot (m(H) - 1)^{(2i-2)/2} \cdot (m(H) - 1)$$
$$= (R(H, H) - 1) \cdot (m(H) - 1)^{(2i)/2}$$

vertices containing no rainbow triangle and no monochromatic copy of H.

Finally suppose $2i + 1 = k$. In this case, construct G_k by making $\chi(H) - 1$ copies of G_{2i} and inserting all edges between the copies in color k. Any subgraph of the graph induced on the edges of color k has chromatic number at most $\chi(H) - 1$ so there is no copy of H in color k. This means that the resulting graph G_k is a k-coloring of K_{n_k} on

$$n_k = (\chi(H) - 1) \cdot (R(H, H) - 1) \cdot (m(H) - 1)^{(2i-2)/2}$$

vertices containing no rainbow triangle and no monochromatic copy of H, completing the proof of Proposition 3.2.   □

In light of Propositions 3.1 and 3.2, the following result by Gyárfás et al. should come at no surprise.

Theorem 3.1 ([3]) *Let H be a fixed graph with no isolated vertex. If H is bipartite and not a star, then $gr_k(K_3 : H)$ is linear in k. If H is not bipartite, then $gr_k(K_3 : H)$ is exponential in k.*

For the case where H is a star, see Theorem 3.7 below. The following proof is adapted from the proof by Gyárfás et al. [3].

Proof The lower bounds for Theorem 3.1 follow from Propositions 3.1 and 3.2. For the upper bound, let $R = R(H, H) - 1$.

First, suppose that H is bipartite. Let

$$n = R \cdot [(\ell(H) - 1)k + 2] \cdot (\ell(H) - 1)$$

and let K be a k-coloring of K_n containing no rainbow triangle. Suppose, for a contradiction, that K also contains no monochromatic copy of H. Since K contains no rainbow triangle, by Theorem 2.1, there is a Gallai partition of $V(K)$. If this partition has at least $R + 1$ parts, then since $R = R(H, H) - 1$, there is a monochromatic copy of H among the edges between the parts of the Gallai partition, a contradiction. Thus, there are at most R parts in the Gallai partition. By the pigeonhole principle, at least one part of the Gallai partition has order at least

$$n/R = [(\ell(H) - 1)k + 2] \cdot (\ell(H) - 1)$$

so let G_1 denote the subgraph of K induced on the vertices of such a (large) part. Note that by the definition of the Gallai partition, each vertex in $V(K) \setminus V(G_1)$ has all edges to G_1 in a single color so assign this color to each vertex of $V(K) - V(G_1)$. Also by the definition of the Gallai partition, there are only two colors assigned to these vertices. If $|V(K)| - |V(G_1)| \geq 2\ell(H) - 1$, then by the pigeonhole principle, there is a set G_1^* of at least $\ell(H)$ vertices all assigned the same color. Then for any choice G_1' of $b(H)$ vertices with $G_1' \subseteq G_1$, the bipartite subgraph induced on $G_1^* \cup G_1'$ induces a monochromatic complete bipartite graph, which contains a monochromatic copy of H. Thus, we may assume that $|V(K)| - |V(G_1)| \leq 2\ell(H) - 2$ so

$$|V(G_1)| \geq n - 2(\ell(H) - 1) = [R \cdot [(\ell(H) - 1)k + 2] - 2] \cdot (\ell(H) - 1).$$

Select a vertex $v_1 \in V(K) \setminus V(G_1)$ and recall that v_1 has all edges to G_1 in a single color. Given such a subgraph G_i for $i \geq 1$, this reduction process can be repeated to produce another subgraph G_{i+1} of order at least $|G_i| - 2(\ell(H) - 1)$ along with a vertex $v_{i+1} \in V(G_i) \setminus V(G_{i+1})$ with all edges to G_{i+1} in a single color. In particular, after $t = (\ell(H) - 1)k + 1$ iterations, the sequence of vertices $v_1, v_2, \ldots, v_t$ has the property that each vertex has all edges to the vertices that follow in a single color. More specifically, given a fixed index i, for all j with $i \leq j \leq t$, all edges of the form $v_i v_j$ have a single color. Furthermore, we get $|G_t| \geq 2\ell(H) - 1 > \ell(H)$. By the pigeonhole principle, there is a set $Z \subseteq \{v_1, v_2, \ldots, v_t\}$ with $|Z| \geq \ell(H)$ such that all vertices in Z have a single color on their incident edges to G_t. Then $Z \cup G_t$

is a monochromatic complete bipartite graph which is large enough to contain a monochromatic copy of H, for a contradiction.

Finally, suppose H is not bipartite. Let

$$n = R^{k(|H|-1)+1}$$

and let K be a k-coloring of K_n containing no rainbow triangle. Suppose, for a contradiction, that K also contains no monochromatic copy of H. As in the previous case, if a Gallai partition of K has at least $R + 1$ parts, then there is a monochromatic copy of H, so there must be at most R parts in any Gallai partition of K. By the pigeonhole principle, at least one part of the Gallai partition has order at least

$$n/R = R^{k(|H|-1)}.$$

Let G_1 be the subgraph of K induced on the vertices of such a (large) part and let v_1 be an arbitrary vertex not in G_1. By the definition of the Gallai partition, all edges from v_1 to G_1 have a single color. Given a subgraph G_i, this reduction process can be repeated to produce another subgraph G_{i+1} of order at least $|G_i|/R$ along with a vertex v_{i+1} with all edges to G_{i+1} in a single color. In particular, after $t = k(\ell(H) - 1) + 1$ iterations, we have $|G_t| \geq 1$ and the sequence of vertices $v_1, v_2, \ldots, v_t$ has the property that given a fixed index i, for all j with $i \leq j \leq t$, all edges of the form $v_i v_j$ have a single color. By the pigeonhole principle, since $t = k(|H| - 1) + 1$ and there are only k colors available, there is a subsequence of vertices $v_{i_1}, v_{i_2}, \ldots, v_{i_{|H|}}$ which induce a monochromatic complete graph. This monochromatic copy of $K_{|H|}$ certainly contains a monochromatic copy of H, for a contradiction. $\square$

In the rest of this section, we gather several useful tools and techniques, lemmas and facts, that are commonly employed in the proofs of upper bounds on Gallai-Ramsey numbers. The first such result eliminates a small case from arguments by using a simple assumption of minimality of the Gallai partition. This concept has appeared in several publications including [3–8] to name a few.

Lemma 3.1 *Given a Gallai colored complete graph G, if a Gallai partition of G is chosen so that the number of parts in the partition is minimum, then there will never be 3 parts in the partition.*

Proof Suppose $t = 3$ is the number of parts in a Gallai partition of G which is chosen to have the minimum number of parts. Then the reduced graph is a 2-colored triangle. Since two edges of this triangle must have the same color, there is a bipartition of the reduced graph, and therefore of G, with all one color on the edges in between the parts, contradicting the minimality of t. See Fig. 3.3. $\square$

Next a result by Hall et al. showing that the colors appearing between parts of a Gallai partition actually each appears at every part. Oftentimes, this lemma is used to show that the degree, restricted to each color appearing between parts of a Gallai partition, of every vertex in the reduced graph is at least 1.

Fig. 3.3 When $t = 3$

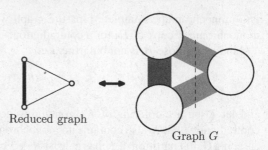

Reduced graph

Graph G

Lemma 3.2 ([9]) *If G is a Gallai colored complete graph, then G has a Gallai partition such that for each color i appearing on edges between the parts, the subgraph of the reduced graph induced by color i is connected. In particular, a Gallai partition chosen to minimize the number of parts in the partition has this property.*

Proof Suppose t is the number of parts in a Gallai partition of G which is chosen to have the minimum number of parts. Assume, for a contradiction, that there is a color that induces a disconnected graph in the reduced graph. Let red be this disconnected color and let blue be the other color appearing on edges of the reduced graph. Then the components of the red subgraph of the reduced graph partition the reduced graph (and therefore the graph) in such a way that all edges between parts of this new partition have blue edges between them. This provides a Gallai partition of G with fewer than t parts, a contradiction to the minimality of t. □

When the desired monochromatic subgraph is bipartite, there is additional information available about the Gallai partition provided by the following result from Wu et al.

Lemma 3.3 ([1]) *Given positive integers ℓ, m, n where $\ell \le m$ and $n \ge 3m - 2$, let $H = K_{\ell,m}$ be a complete bipartite graph and let G be a Gallai coloring of K_n with no monochromatic copy of H. In any Gallai partition of G, any largest part of the partition has order either at most $\ell - 1$ or at least $n - 2\ell + 2$.*

We include a modification of the proof by Wu et al. [1].

Proof Let H and G be as given and let H_1 be a largest part of a Gallai partition of G. Suppose, for a contradiction, that $\ell \le |H_1| \le n - 2\ell + 1$.

First suppose that $|H_1| \le n - 2m + 1$. Since H_1 is a part of a Gallai partition, say with red and blue appearing on the edges between parts, every other vertex of G has either all red edges or all blue edges to H_1. By the pigeonhole principle, there must be a set S of at least m vertices outside H_1 (the union of some other parts of the Gallai partition) with all one color on edges to H_1, say red. Since $|H_1| \ge \ell$, the (red) complete bipartite graph $S \cup H$ contains a red copy of $K_{\ell,m}$, a contradiction.

Finally suppose that $n - 2m + 2 \le |H_1| \le n - 2\ell + 1$. This means that there are at least $2\ell - 1$ vertices in $G \setminus H_1$. Since $n \ge 3m - 2$, we get $|H_1| \ge n - 2m + 2 \ge m$. By the pigeonhole principle, there is a set S of at least ℓ vertices in $G \setminus H_1$ (again the union of some other parts of the Gallai partition) with the property that all edges between S and H_1 have the same color. Then $S \cup H_1$ induces a monochromatic graph containing $K_{\ell,m}$, a contradiction completing the proof of Lemma 3.3. $\qquad\square$

Under extra restrictions on the graph H and its 2-color Ramsey number, we get even more helpful information.

Theorem 3.2 ([1]) *Given a bipartite graph H and a positive integer R with $R \geq$ $\max\{R(H, H), 3\ell(H) - 2\}$, if every Gallai coloring of K_R using 3 colors, in which all parts of a Gallai partition have order at most $s(H) - 1$, contains a monochromatic copy of H, then*

$$gr_k(K_3 : H) \leq R + (s(H) - 1)(k - 2).$$

An alternative statement of this result is as follows: Given a bipartite graph H and a given positive integer $R \geq \max\{R(H, H), 3\ell(H) - 2\}$, if we intend to prove that

$$gr_k(K_3 : H) \leq R + (s(H) - 1)(k - 2),$$

then it suffices to prove that every Gallai coloring of K_R using only 3 colors, in which all parts of the Gallai partition have order at most $s(H) - 1$, contains a monochromatic copy of H.

We include a modified version of the proof by Wu et al. [1].

Proof Let H be the given bipartite graph with $a = s(H)$ and $b = |H| - a$ so $H \subseteq K_{a,b}$. Assume that every Gallai coloring of K_R using only 3 colors, in which parts of the Gallai partition have order at most $s(H) - 1$, contains a monochromatic copy of H. Then we will show that $gr_k(K_3 : H) \leq R + (s(H) - 1)(k - 2)$.

Let G be an arbitrary Gallai coloring of K_n where $n = R + (a - 1)(k - 2)$ and suppose G contains no monochromatic copy of H. If there is a small set of vertices T_0 with all one color c on their incident edges, then we must have $|T_0| \leq s(H)$ to avoid creating a monochromatic copy of H in color c.

More specifically, define a (possibly empty) set T of vertices $\{v_1, v_2, \ldots, v_{|T|}\}$ to be a largest set of vertices in $V(G)$ with the following properties:

- each vertex v_i has all edges to $G \setminus T$ in a single color C_i, and
- there are at least b vertices in $G \setminus T$.

We first observe that $|T| \leq (a - 1)k$ since otherwise, if $|T| \geq (a - 1)k + 1$, then there would exist a color, say c, such that at least a vertices among the first $(a - 1)k + 1$ vertices of T have all edges to $G \setminus T$ in color c. This yields a monochromatic copy of $K_{a,b}$ in color c, which contains the desired monochromatic copy of H. In fact, for each color c, this same argument shows that the set of vertices $T_c \subseteq T$ with all incident edges to $G \setminus T$ in color c must have $|T_c| \leq a - 1$.

By Theorem 2.1, there is a Gallai partition of $G \setminus T$, say using colors red and blue. By Lemma 3.3, the largest part of any Gallai partition of $G \setminus T$, say H_1, has order either at least $|G \setminus T| - 2a + 2$ or at most $a - 1$. First suppose H_1 has order at least $|G \setminus T| - 2a + 2$. Then all the vertices of $(G \setminus T) \setminus H_1$ can be added to T since each such vertex has only one color on its incident edges to H_1, contradicting the maximality of $|T|$. We may therefore assume that $|H_1| \leq a - 1$, and therefore all parts of this Gallai partition have order at most $a - 1$.

Let A and B be the (possibly empty) sets of vertices in T with all edges from A to $G \setminus T$ being red and all edges from B to $G \setminus T$ being blue. Define the graph $G' = (G \setminus T) \cup (A \cup B)$. Note that $|G'| \geq R$. From G', construct a new graph G'' on the same vertex set by changing the colors of

- all edges within the parts of the Gallai partition of $G \setminus T$,
- all edges within A and
- all edges within B

to green. Then G'' is a 3-colored complete graph with $|G''| \geq R$. Furthermore, G'' is a Gallai coloring and there is a Gallai partition of G'' using colors red and blue with each part having order at most $a - 1$.

By construction, if G'' contains a monochromatic copy of H then G contains a monochromatic copy of H (since there can certainly be no green copy of H).   □

3.2   Sharp Results for Bipartite Graphs

We first present some general containment results to preface the section.

Given two integers $h \geq 2$ and $b \geq 0$, the *broom* graph $B_{h,b}$ is a tree constructed by identifying an end vertex of a path P_h with the center vertex of a star S_b on b edges. The path portion of the broom is called the *handle* while the star portion is called the *bristles*. See Fig. 3.4 for an example.

Within 2-colorings of complete graphs, two types of monochromatic spanning trees are known to exist.

Theorem 3.3 ([10]) *In every 2-coloring of a complete graph, there exists a spanning monochromatic broom.*

Theorem 3.4 ([11]) *In every 2-coloring of a complete graph, there exists a monochromatic spanning tree with height at most 2.*

Using Theorem 2.1, Gyárfás and Simonyi prove the following results, which extend Theorems 3.3 and 3.4, respectively, from 2-colorings to Gallai colorings.

Fig. 3.4 A broom

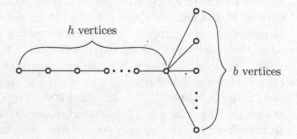

h vertices

b vertices

Fig. 3.5 Lower bound example for monochromatic stars

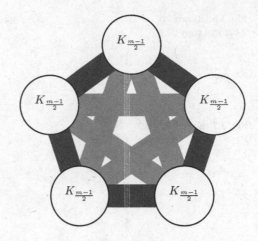

Theorem 3.5 ([7]) *In every Gallai coloring of a complete graph, there exists a spanning monochromatic broom.*

Theorem 3.6 ([7]) *In every Gallai coloring of a complete graph, there exists a monochromatic spanning tree with height at most 2.*

We next present the complete solution for monochromatic stars by Gyárfás and Simonyi.

Theorem 3.7 ([7]) *For $m \geq 2$ and $k \geq 3$,*

$$gr_k(K_3 : S_m) = \begin{cases} \frac{5m-6}{2} & \text{if } m \text{ is even,} \\ \frac{5m-3}{2} & \text{if } m \text{ is odd.} \end{cases}$$

This proof is based on the proof presented in [7].

Proof Let $m \geq 2$ and $k \geq 3$. For simplicity, we assume appropriate divisibility throughout this proof. For the lower bound, construct G from 5 copies of $K_{(m-1)/2}$ each colored entirely in color 1. The edges between these copies are then colored so that the copies form a blow-up of the 2-colored K_5 with no monochromatic triangle, each color inducing a C_5. See Fig. 3.5. This graph contains no monochromatic star and has order $\frac{5m-5}{2}$.

For the upper bound, let $n \geq \frac{5m-3}{2}$ and let G be a Gallai coloring of K_n. Consider a Gallai partition of G into the smallest number t of parts $H_1, H_2, \ldots, H_t$, say with $|H_1| \leq |H_2| \leq \cdots \leq |H_t|$. Suppose red and blue are the colors used on the edges in this Gallai partition.

First suppose $t \leq 3$. By Lemma 3.1, this means we may assume $t = 2$. Then $|H_2| \geq \frac{n}{2}$ so any vertex in H_1 has at least $\frac{n}{2}$ edges (to H_2) in a single color, creating the desired monochromatic star.

Fig. 3.6 Blown-up
2-colored copy of K_4

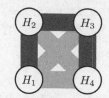

Fig. 3.7 Structure when
$t \geq 5$

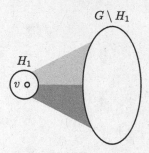

Next suppose $t = 4$. Then, by minimality of t, each vertex of the reduced graph has exactly two incident edges in one color and one incident edge in the other color. See Fig. 3.6 for an example of this structure. This means that a vertex $v \in H_1$ has all edges in one color to two of H_2, H_3 or H_4, say H_i and H_j for some choice of i and j with $2 \leq i < j \leq 4$. If $|H_i| + |H_j| \geq \frac{2n}{5}$, the proof is complete, so suppose not. This means that $|H_i| + |H_j| < \frac{2n}{5}$. Since $|H_1| \leq |H_2| \leq |H_3| \leq |H_4|$, this means that $|H_1| < \frac{n}{5}$. With $n = \sum_{i=1}^{4} |H_i|$, this implies $|H_4| > \frac{2n}{5}$. Since every vertex in $G \setminus H_1$ has all one color on its edges to H_1, each such vertex is the center of a desired monochromatic star.

Finally suppose $t \geq 5$. Then $|H_1| \leq \frac{n}{5}$, so let $v \in H_1$. All edges from v to $G \setminus H_1$ are either red or blue (see Fig. 3.7) so with $|G \setminus H_1| \geq \frac{4n}{5}$, any such vertex v must have at least $\frac{2n}{5}$ incident edges in a single color (red or blue), making v the center of a desired monochromatic star and completing the proof of Theorem 3.7. $\qquad\square$

Next a sharp result by Faudree et al. for the path on 4 vertices.

Theorem 3.8 ([6]) *For all $k \geq 1$, we have*

$$gr_k(K_3 : P_4) = k + 3.$$

This proof is slightly expanded from the original proof in [6].

Proof The lower bound for $k \geq 2$ follows from Proposition 3.1 since $R(P_4, P_4) = 5$. For $k = 1$, the lower bound follows from considering a 1-colored K_3 which certainly cannot contain a monochromatic P_4.

For the upper bound, for a contradiction, suppose G is a k-colored K_n with $n = k + 3$ containing no rainbow triangle and no monochromatic P_4. We prove this result

Fig. 3.8 The structure of G,
all pictured edges are red

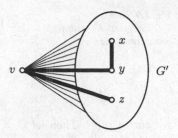

by induction on the number of colors k, where the base of this induction, when $k = 1$, follows since any 1-coloring of K_4 contains a monochromatic P_4.

By Theorem 3.5, there is a spanning monochromatic broom $B_{h,b}$ in G, say in red. If $h \geq 4$ or $h \geq 3$ and $b \geq 1$, then this broom contains a red P_4, for a contradiction. Thus, since $n = k + 3 \geq 5$, this broom must be a spanning red star, or rather $h = 2$ and $b = n - 2$. Let v be the center of this star and let $G' = G \setminus \{v\}$. Within G', if there is a red edge, say xy, then for any other vertex $z \in G'$, the path $xyvz$ is a red P_4, a contradiction. See Fig. 3.8 where all pictured edges are red and the bold path is the constructed P_4. Thus, there is no red edge in G', so by induction on k applied within G', the desired result holds. □

The Gallai-Ramsey numbers for several small paths are known precisely while a general bound is presented in Sect. 3.4 (see Theorem 3.36).

Theorem 3.9 ([6, 12])

- $gr_k(K_3 : P_4) = k + 3$,
- $gr_k(K_3 : P_5) = k + 4$,
- $gr_k(K_3 : P_6) = 2k + 4$,
- $gr_k(K_3 : P_7) = 2k + 5$,
- $gr_k(K_3 : P_8) = 3k + 5$.

As another example of a standard proof strategy, we present the Gallai-Ramsey number for C_4. This proof, in particular, is indicative of the strategy employed in many of the proofs in that it almost uses induction on the number of colors but such a direct induction is not quite possible so a slight modification is needed.

Theorem 3.10 ([6]) *For all $k \geq 2$, we have*

$$gr_k(K_3 : C_4) = k + 4.$$

This proof is adapted and expanded from the original proof by Faudree et al. in [6].

Proof The lower bound follows from Proposition 3.1 since $R(C_4, C_4) = 6$. In order to complete the proof of the upper bound, we actually prove a slightly stronger result but in order to state this result, we need some definitions. Given a k-colored complete graph G, let G_i denote the subgraph of G using precisely all the edges of color i

Fig. 3.9 The two
2-colorings of K_5 with no
monochromatic C_4

 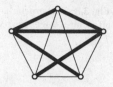

for each $i \in [k]$. Then let $k' = k'(G)$ be the number of these subgraphs G_i such
that $\Delta(G_i) \geq 2$. This means that k' is the number of colors used in G for which
there is a vertex with at least two incident edges in that color. Certainly $k' \leq k$. Let
$gr'_{k'}(K_3 : C_4)$ denote the minimum integer n such that any coloring of K_n in which
at most k' colors induce a subgraph with maximum degree at least 2 must contain
either a rainbow triangle or a monochromatic copy of C_4. Then it suffices to prove
that

$$gr'_{k'}(K_3 : C_4) = k' + 4.$$

Note that since $k' \leq k$, this implies the desired upper bound. Also this has no effect
on the lower bound since our construction of G_k has $k'(G_k) = k$.

Suppose G is a k-coloring of $K_{k'+4}$ containing no rainbow triangle in which at
most k' colors induce a subgraph with maximum degree at least 2 and suppose, for
a contradiction, that G also contains no monochromatic C_4. For induction, suppose
that the result holds for smaller values of k', that is, if G is a k-coloring of $K_{k'+3}$ with
$k'(G) = k' - 1$, then G contains either a rainbow triangle or a monochromatic C_4.
If $k' = 0$, every color induces a matching and so every set of three vertices induces
a rainbow triangle, a contradiction. If $k' = 1$, then G is a coloring of K_5 in which
at most one color (say color 1) induces a subgraph with a vertex of degree at least
2. Here we consider color 1 as one color and all other colors as color 0. If there
is a cycle in color 0, then using some chord of the cycle, we can make a smaller
properly colored cycle and repeat until we arrive at a rainbow triangle. Recall that
$R(C_4, C_4) = 6$ and there are precisely two sharpness examples on 5 vertices. In one
example, both colors induce a C_5 and in the other, both colors induce a triangle with
two disjoint pendant edges. See Fig. 3.9. Since there is a cycle in both colors in both
examples, we find a rainbow triangle in both examples for a contradiction.

By Theorem 2.1, there is a Gallai partition of the vertices (meaning that on all
edges between each pair of parts, there is only one color, and on edges between
the parts in general, there are only two colors). Let $H_1, H_2, \ldots, H_t$ be the parts of
this Gallai partition, say with $|H_1| \geq |H_2| \geq \cdots \geq |H_t|$. Since this is a nontrivial
partition, we have $t \geq 2$.

If $|H_{t-1}| \geq 2$, then each of H_{t-1} and H_t contains at least two vertices, say $u, v \in$
H_{t-1} and $x, y \in H_t$. Since all edges between H_{t-1} and H_t have a single color, the
cycle $uxvyu$ is a monochromatic C_4 for a contradiction. We may therefore assume
that $|H_{t-1}| = 1$.

If $|H_t| = 1$, then all parts have order 1 and G is simply a 2-coloring of $K_{k'+4}$.
Since $R(C_4, C_4) = 6$, we see that $k' + 4 \leq 5$ so $k' \leq 1$ and this is precisely the base
case described above.

Finally if $|H_t| \geq 2$ and $|H_{t-1}| = 1$, we recall, from the structure of the Gallai partition, that each vertex in $G \setminus H_t$ has all edges in a single color to H_1, and these edges come from only two colors. If there are at least three vertices in $G \setminus H_t$, then two of them must have the same color, say red, on all edges to H_t. These two vertices, along with two vertices from H_t induce a red C_4 in G for a contradiction, meaning that there can be only two vertices in $G \setminus H_t$, so $t \leq 3$.

Let v be one of the vertices in $G \setminus H_t$, say with all red edges to H_t. Notice that there can be no vertex in H_t with two incident red edges. Since each vertex in $G \setminus H_t$ has a distinct color on all edges to H_t, we see that $k'(H_t) = k' - (t - 1)$, completing the proof by induction on k'. □

Only a few sharp results are known for even cycles.

Theorem 3.11 ([6, 13, 14])

- $gr_k(K_3 : C_4) = k + 4$,
- $gr_k(K_3 : C_6) = 2k + 4$,
- $gr_k(K_3 : C_8) = 3k + 5$.

See Theorem 3.30 for general bounds on even cycles.

Sharp results for several small trees are also known, see the tables in Sect. 5.1 for a complete list of known sharp results to the best of our knowledge at the time of writing this. Please refer to [15] by Fujita et al. for a dynamically updated survey of results of this form.

3.3 Sharp Results for Non-Bipartite Graphs

We break this section into subsections for certain classes of graphs.

3.3.1 Odd Cycles

Finding either a rainbow triangle or a monochromatic triangle has been well studied. Indeed, the following result was considered from three different perspectives. First, in [5], Chung and Graham considered colorings of complete graphs in which every triangle uses precisely 2 colors. From a different perspective, by forbidding a monochromatic triangle, Axenovich and Iverson [4] forced the existence of rainbow triangles. Finally from yet another perspective, Gyárfás et al. [3] used forbidden rainbow triangles and the structure provided by Theorem 2.1 to force monochromatic triangles.

Theorem 3.12 ([3–5]) *For $k \geq 1$,*

$$gr_k(K_3 : K_3) = \begin{cases} 5^{k/2} + 1 & \text{if } k \text{ is even, or} \\ 2 \cdot 5^{(k-1)/2} + 1 & \text{if } k \text{ is odd.} \end{cases}$$

We include a proof adapted from ideas in [3] by Gyárfás et al.

Proof The lower bound follows from Proposition 3.2. The upper bound is proven by induction on k. The case where $k = 1$ follows trivially by considering a 1-colored triangle while the case where $k = 2$ follows from the fact that $R(K_3, K_3) = 6$. Thus, suppose $k \geq 3$.

Let G be a Gallai coloring using k colors of a complete graph on n vertices where

$$n = n_k = \begin{cases} 5^{k/2} + 1 & \text{if } k \text{ is even, or} \\ 2 \cdot 5^{(k-1)/2} + 1 & \text{if } k \text{ is odd.} \end{cases}$$

By Theorem 2.1, there is a Gallai partition of G, say using colors red and blue. Consider such a Gallai partition with the smallest number of parts, say t. Note that $t \leq 5$ since $R(K_3, K_3) = 6$ and the reduced graph (the graph induced on any selection of one vertex from each part) is a 2-colored K_t. Let $H_1, H_2, \ldots, H_t$ be the parts of the Gallai partition.

First suppose $2 \leq t \leq 3$. If $t \leq 3$, then by Lemma 3.1, we may assume $t = 2$. Suppose red is the color of the edges between the two parts. Then, to avoid a red triangle, there can be no red edges within either part. Since a color is missing within each part H_i, apply induction on k within H_i. This means that

$$\begin{aligned} |G| &= |H_1| + |H_2| \\ &\leq 2[n_{k-1} - 1] \\ &= 2 \cdot \begin{cases} 5^{(k-1)/2} & \text{if } k \text{ is odd,} \\ 2 \cdot 5^{(k-2)/2} & \text{if } k \text{ is even} \end{cases} \\ &= \begin{cases} 2 \cdot 5^{(k-1)/2} & \text{if } k \text{ is odd,} \\ 4 \cdot 5^{(k-2)/2} & \text{if } k \text{ is even} \end{cases} \\ &< n, \end{aligned}$$

for a contradiction.

Thus, we may assume that $4 \leq t \leq 5$. If there is a part H_i such that the corresponding vertex in the reduced graph is incident to only edges of one color, then there is again a bipartition of the vertices of the reduced graph, and therefore the vertices of G, with all one color on edges between the parts. Thus, by minimality of t, each of the vertices corresponding to parts H_i has incident edges in both red and blue in the reduced graph. Then, to avoid creating a monochromatic triangle, each part H_i contains no red or blue edges, so each part is missing at least two colors. We may therefore apply induction on k within each part H_i. This means that

$$|G| = \sum_{i=1}^{t} |H_i|$$
$$\leq 5[n_{k-2} - 1]$$
$$= 5 \cdot \begin{cases} 5^{(k-2)/2} & \text{if } k \text{ is even,} \\ 2 \cdot 5^{(k-3)/2} & \text{if } k \text{ is odd} \end{cases}$$
$$= \begin{cases} 5^{k/2} & \text{if } k \text{ is even,} \\ 2 \cdot 5^{(k-1)/2} & \text{if } k \text{ is odd} \end{cases}$$
$$= n - 1$$
$$< n,$$

again a contradiction, completing the proof. $\square$

Several other specific cases of odd cycles have also been shown precisely by a variety of authors.

Theorem 3.13 ([13, 16–18])

- $gr_k(K_3 : C_5) = 2^{k+1} + 1$,
- $gr_k(K_3 : C_7) = 3 \cdot 2^k + 1$,
- $gr_k(K_3 : C_9) = 4 \cdot 2^k + 1$,
- $gr_k(K_3 : C_{11}) = 5 \cdot 2^k + 1$,
- $gr_k(K_3 : C_{13}) = 6 \cdot 2^k + 1$,
- $gr_k(K_3 : C_{15}) = 7 \cdot 2^k + 1$.

Finally, the general result for all odd cycles was settled by Wang et al. [19] and independently by Zhang et al. [20].

Theorem 3.14 ([19, 20]) *For integers $\ell \geq 3$ and $k \geq 1$, we have*

$$gr_k(K_3 : C_{2\ell+1}) = \ell \cdot 2^k + 1.$$

The proof in [19] includes the following main tool, which applies much more generally than just to odd cycles.

Lemma 3.4 *Let $k \geq 3$, H be a graph with $|H| = m$, and let G be a Gallai coloring of the complete graph K_n containing no monochromatic copy of H. If $V(G) = A \cup B_1 \cup B_2 \cup \cdots \cup B_{k-1}$ where $G[A]$ uses at most k colors (say from $[k]$), $|B_i| \leq m - 1$ for all i, and all edges between A and B_i have color i, then $n \leq gr_k(K_3 : H) - 1$.*

Note that this lemma uses the assumed structure to produce a bound on $|G|$ even if the colored complete graph G itself uses more than k colors. Other colors (with index greater than k) may appear within the sets B_i. We include the straightforward proof from [19] for completeness.

Proof For $i \neq j$, all edges between B_i and B_j must have either color i or color j to avoid a rainbow triangle. Since $|B_i| \leq m - 1 < |H|$, changing all edges within B_i that are not a color in $[k - 1]$ to color k cannot create a monochromatic copy of H. The result of this modification is a k-colored copy of K_n with no rainbow triangle and no monochromatic copy of H so $n \leq gr_k(K_3 : H) - 1$. $\qquad\square$

The lemma to follow is an easy consequence of this classical result of Dirac.

Theorem 3.15 ([21]) *Let G be a graph of order $n \geq 3$. If the minimum degree of G satisfies $\delta(G) \geq \frac{n}{2}$, then G is hamiltonian.*

Lemma 3.5 *If there are at least $2\ell + 1$ vertices in a Gallai colored complete graph G with only one color appearing on edges between parts of the Gallai partition and all parts of order at most ℓ, then G contains a monochromatic copy of $C_{2\ell+1}$.*

Proof Suppose blue is the color of the edges between parts. Since each part of the Gallai partition has order at most ℓ, every vertex is incident with at least $2\ell - (\ell - 1) \geq \ell + 1$ blue edges. By Theorem 3.15, G contains a blue copy of $C_{2\ell+1}$. $\qquad\square$

Next a helpful result about the existence of paths as subgraphs of graphs. This will be applied within monochromatic subgraphs to produce monochromatic paths of appropriate lengths.

Theorem 3.16 ([22]) *Let G be a graph on n vertices and let $k \geq 2$ be an integer. If the number of edges in G satisfies $e(G) > \frac{k-1}{2}n$, then G contains a path with k edges (i.e. a copy of P_{k+1}).*

We use Theorem 3.16 to prove the following colored version. Here we let $d_R(v)$ and $d_B(v)$ denote the red and blue degrees of the vertex v, respectively, that is, the number of edges incident to v that have color red and blue, respectively.

Lemma 3.6 *Let G be a graph of order n with edges colored by red and blue. If for every vertex $v \in V(G)$ and for some nonnegative integers a and b with $a + b \geq 3$, we have $d(v) \geq a + b - 3$, then G contains either a red copy of P_a or a blue copy of P_b.*

Proof Let $\bar{d}(G)$, $\bar{d}_R(G)$ and $\bar{d}_B(G)$ denote the average degree, average red degree and average blue degree of G, respectively. Then $\bar{d}(G) = \bar{d}_R(G) + \bar{d}_B(G)$. Since $d_R(v) + d_B(v) \geq a + b - 3$ for every vertex $v \in V(G)$, we have $\bar{d}_R(G) + \bar{d}_B(G) \geq a + b - 3$ so either $\bar{d}_R(G) > a - 2$ or $\bar{d}_B(G) > b - 2$. If $\bar{d}_R(G) > a - 2$, then there are more than $\frac{n(a-2)}{2}$ red edges in G, so by Theorem 3.16, G contains a red copy of P_a. On the other hand, if $\bar{d}_B(G) > b - 2$, then there are more than $\frac{n(b-2)}{2}$ blue edges in G, so by Theorem 3.16, G contains a blue copy of P_b, completing the proof. $\qquad\square$

Finally, we state some other known results that will be used in the proof.

Theorem 3.17 ([23–25])

$$R(C_m, C_n) = \begin{cases} 2n - 1 \\ \quad for\, 3 \leq m \leq n,\, m\, and\, n\, odd,\, (m, n) \neq (3, 3), \\ n - 1 + m/2 \\ \quad for\, 4 \leq m \leq n,\, m\, and\, n\, even,\, (m, n) \neq (4, 4), \\ \max\{n - 1 + m/2,\, 2m - 1\} \\ \quad for\, 4 \leq m < n,\, m\, even\, and\, n\, odd. \end{cases}$$

Theorem 3.18 ([9, 13]) *Given integers $n \geq 2$ and $k \geq 1$,*

$$(n - 1)k + n + 1 \leq gr_k(K_3 : C_{2n}) \leq (n - 1)k + 3n.$$

In particular, we mostly use an immediate corollary of Theorem 3.18, namely, that $gr_3(K_3 : C_{2\ell-2}) \leq 6\ell - 9$.

Proof (*of Theorem* 3.14 *from* [19]) This proof is by induction on k. Throughout the proof, we will eliminate edges of some colors from subgraphs and then apply induction on the number of colors available for use on the edges within those subgraphs to provide bounds on the total number of vertices. Since the cases when $k = 1$ and $k = 2$ are either trivial or follow immediately from Theorem 3.17, respectively, we assume that $k \geq 3$. For a contradiction, suppose that $\ell \geq 3$ and let G be a Gallai colored complete graph K_n containing no monochromatic copy of $C_{2\ell+1}$ with

$$n = \ell \cdot 2^k + 1.$$

The goal is to arrive at a contradiction.

Let T be a maximal set of vertices $T = T_1 \cup T_2 \cup \cdots \cup T_k$ where each subset T_i has all edges to $G \setminus T$ in color i and $|G \setminus T| \geq \ell + 1$ constructed iteratively by adding at most 2ℓ vertices at a time, with at most ℓ vertices being added to each of two sets T_i at a time. The set T is initialized with an empty set and vertices (or small sets of at most ℓ vertices at a time) are added to T if there exists a vertex (or small set) with all one color on edges to the rest of the graph. We first claim that $|T_i|$ is small for each value of i.

Claim For each i with $1 \leq i \leq k$, we have $|T_i| \leq \ell$ and furthermore, $T_i = \emptyset$ for some value of i.

Proof By the iterative definition of T, we may assume that this is the first step in the iterative construction where T violates one of these assumptions. That is, assume that $|T_i| \leq 2\ell$ for all i and either

- $T_\alpha = \emptyset$ for some α and at most two sets T_i and T_j have $|T_i|, |T_j| > \ell$, or
- no set is empty and at most one set T_i has $|T_i| > \ell$.

In either of these cases, we have $|T| \leq (k+1)\ell$ since

$$|T| \leq \begin{cases} (k-3)\ell + 2(2\ell), \\ (k-1)\ell + 2\ell. \end{cases}$$

We first show that $T_i = \emptyset$ for some i so suppose the latter item above. If $k \geq 4$, then

$$|G \setminus T| \geq [\ell 2^k + 1] - [(k+1)\ell] \geq (\ell - 1)k + 3\ell.$$

By Theorem 3.18, there is a monochromatic copy of $C_{2\ell}$ contained within $G \setminus T$. Since $T_i \neq \emptyset$ for all i, this cycle can easily be extended to a monochromatic copy of $C_{2\ell+1}$, for a contradiction. We may therefore assume $k = 3$.

If some set T_i has $|T_i| \geq \ell + 1$, then $G \setminus T$ contains no edges of color i since otherwise if such an edge $e = u_1 u_2$ exists, then letting $\{u_3, u_4, u_{\ell+1}\}$ be any set of $\ell - 1$ vertices in $(G \setminus T) \setminus \{u_1, u_2\}$ and $\{v_1, v_2, \ldots, v_\ell\}$ be any set of ℓ vertices in T_i, then the cycle $u_1 u_2 v_1 u_3 v_2 u_4 \ldots u_{\ell+1} v_\ell u_1$ would be a copy of $C_{2\ell+1}$ in color i. Thus

$$|G \setminus T| \geq (\ell - 1)3 + 3\ell = 6\ell - 3 > 4\ell + 1$$

so, by Theorem 3.17, there is a monochromatic copy of $C_{2\ell+1}$ in one of the other colors. We may therefore assume that $|T_i| \leq \ell$ for all $i \in \{1, 2, 3\}$ and so $|G \setminus T| \geq 5\ell + 1$.

Within $G' = G \setminus T$, consider a Gallai partition and let r be the number of parts in this partition of order at least ℓ and suppose red and blue are the colors appearing on edges between the parts of the partition. Without loss of generality, let color 1 be red and color 2 be blue. If $r \geq 2$, then two of these large parts, say with red edges between, along with any vertex of T_1, produces a red copy of $C_{2\ell+1}$. This means we may immediately assume $r \leq 1$. If $r = 1$ say with H_1 as the large part, then let G_R and G_B be the sets of vertices in $G' \setminus H_1$ with all red and blue (respectively) edges to H_1. If $|G_R \cup T_1| \geq \ell + 1$ or $|G_B \cup T_2| \geq \ell + 1$, then there is a red or blue copy of $C_{2\ell+1}$, respectively, meaning that $|G_R \cup T_1|, |G_B \cup T_2| \leq \ell$. Then G_R and G_B can be added to T to produce a larger set than T with the same properties, contradicting the maximality of T. We may therefore assume $r = 0$. Within G', every vertex has degree at least $|G'| - (\ell - 1)$ when restricted to the red and blue edges. Since $|G'| \geq 5\ell + 1$, we have

$$d(v) \geq 5\ell - (\ell - 1) = 4\ell + 1 > [2\ell] + [2\ell] - 3$$

so, by Lemma 3.6, there must exist either a red path or a blue path of order at least 2ℓ. This path along with a vertex of T with appropriately colored edges produces a monochromatic copy of $C_{2\ell+1}$, a contradiction, completing the proof that $T_i = \emptyset$ for some i.

Next we show that $|T_i| \leq \ell$ for all i. First suppose there is only one set, T_i with $|T_i| > \ell$. Any edge of color i within $G \setminus T$ would produce a monochromatic copy of

$C_{2\ell+1}$ so $G \setminus T$ contains no edge of color i. Then by Lemma 3.4 and induction on k, we have $|G \setminus T_i| \leq gr_{k-1}(K_3 : C_{2\ell+1}) - 1$ so

$$
\begin{aligned}
n &= |T_i| + |G \setminus T_i| \\
&\leq 2\ell + gr_{k-1}(K_3 : C_{2\ell+1}) - 1 \\
&= 2\ell + \ell \cdot 2^{k-1} \\
&< \ell \cdot 2^k + 1 \\
&\leq gr_k(K_3 : C_{2\ell+1}),
\end{aligned}
$$

a contradiction.

Finally suppose there are two sets T_i and T_j with $\ell + 1 \leq |T_i| \leq 2\ell$ and $\ell + 1 \leq |T_j| \leq 2\ell$ so $G \setminus T$ contains no edge of color i or j. Then by Lemma 3.4, we have $|G \setminus (T_i \cup T_j)| \leq gr_{k-2}(K_3 : C_{2\ell+1}) - 1$ so

$$
\begin{aligned}
n &= |T_i| + |T_j| + |G \setminus (T_i \cup T_j)| \\
&\leq 2\ell + 2\ell + gr_{k-2}(K_3 : C_{2\ell+1}) - 1 \\
&= 4\ell + \ell \cdot 2^{k-2} \\
&< \ell \cdot 2^k + 1 \\
&\leq gr_k(K_3 : C_{2\ell+1}),
\end{aligned}
$$

a contradiction, completing the proof of Claim 3.3.1. □

Consider a Gallai partition of $G \setminus T$ with the minimum number of parts t and let $H_1, \ldots, H_t$ be the parts of the partition where $|H_1| \geq |H_2| \geq \cdots \geq |H_t|$, say with $|H_1| = s$. Certainly $s \geq 2$ because otherwise $G \setminus T$ would be a 2-coloring with

$$
|G \setminus T| \geq \ell \cdot 2^k + 1 - (k+1)\ell = \ell(2^k - (k+1)) + 1 \geq 4\ell + 1,
$$

producing the desired monochromatic cycle by Theorem 3.17. Without loss of generality, suppose red (color 1) and blue (color 2) are the two colors appearing on edges between parts in the Gallai partition. Note that the choice of t to be minimum implies that both red and blue are either connected or absent in the reduced graph so in particular, every part has red edges to at least one other part and blue edges to at least one other part (note that if $t = 2$, there would necessarily be only one such color).

Claim We may assume $|H_1| \leq \ell$.

Proof Suppose not, so $|H_1| \geq \ell + 1$. Let r be the number of parts with H_i with $|H_i| \geq \ell + 1$ so $|H_1| \geq |H_2| \geq \cdots \geq |H_r| \geq \ell + 1$ and call these parts *large*. Certainly $r \leq 5 = R(K_3, K_3) - 1$ since any monochromatic triangle in the reduced graph among these large parts would yield a monochromatic copy of $C_{2\ell+1}$. We break the remainder of the proof into cases based on the value of r.

Case 3.1 $r = 5$.

Since there can be no monochromatic triangle in the reduced graph restricted to the 5 large parts, the reduced graph must be the unique 2-coloring of K_5 containing two complimentary copies of C_5. Either one of these cycles yields a monochromatic copy of $C_{2\ell+1}$, completing the proof of this case.

Case 3.2 $r = 4$.

If $t = 4$, then there are no vertices in $(G \setminus T) \setminus (H_1 \cup \cdots \cup H_4)$, so since $|H_i| \geq \ell + 1$, no part can contain any red or blue edges as such an edge would yield a monochromatic copy of $C_{2\ell+1}$. This also means that $T_1 = T_2 = \emptyset$. Define $H_1' = H_1 \cup T_3$, $H_2' = H_2 \cup T_4 \cup T_5 \cup \cdots \cup T_k$ and let $H_i' = H_i$ for $i \geq 3$. Then by induction on k, we have $|H_i'| \leq gr_{k-2}(K_3 : C_{2\ell+1}) - 1$ for $i \in \{3, 4\}$ and by Lemma 3.4 (which is made possible by the fact that $|T_i| \leq 2\ell$) and induction on k, we have $|H_i'| \leq gr_{k-2}(K_3 : C_{2\ell+1}) - 1$ for $i \in \{1, 2\}$. We therefore get that

$$
\begin{aligned}
n &= \sum_{i=1}^{4} |H_i'| \\
&\leq 4[gr_{k-2}(K_3 : C_{2\ell+1}) - 1] \\
&= 4(\ell \cdot 2^{k-2}) \\
&< \ell \cdot 2^k + 1 \\
&\leq gr_k(K_3 : C_{2\ell+1}),
\end{aligned}
$$

a contradiction. This means that $t \geq 5$ so there is at least one vertex v in $(G \setminus T) \setminus (H_1 \cup \cdots \cup H_4)$. The reduced graph restricted to the large parts could either be two complementary copies of P_4 or a matching in one color with all other edges between parts in the other color.

First suppose the reduced graph consists of a red matching, say $H_1 H_2$ and $H_3 H_4$ with all blue edges otherwise in between the parts. In order to avoid creating a red copy of $C_{2\ell+1}$ using edges between H_1 and H_2 along with v, the vertex v must have all blue edges to one of H_1 or H_2 and similarly to one of H_3 or H_4, say H_1 and H_3. Then the blue edges between H_1 and H_3 along with v yield a blue copy of $C_{2\ell+1}$, for a contradiction.

Finally suppose the reduced graph is two complementary copies of P_4, say $H_1 H_2 H_3 H_4$ in red and the remaining edges in blue. To avoid creating a blue copy of $C_{2\ell+1}$ using the blue edges between H_1 and H_4 along with v, the vertex v must have all red edges to either H_1 or H_4, suppose H_1. In order to avoid creating a red copy of $C_{2\ell+1}$ using the red edges between H_1 and H_2 along with v, the vertex v must have all blue edges to H_2. In order to avoid creating a blue copy of $C_{2\ell+1}$ using the blue edges between H_2 and H_4 along with v, the vertex v must have all red edges to H_4. Then $v H_1 H_2 H_3 H_4 v$ induces a red copy of C_5 in the reduced graph, yielding a red copy of $C_{2\ell+1}$ in G for a contradiction, completing the proof in this case.

Case 3.3 $r = 3$.

The cycle among the 3 large parts cannot be monochromatic so suppose, without loss of generality, that the edges from H_2 to H_3 are blue and all other edges between these parts are red. Let A be the set of vertices in $G \setminus (H_1 \cup H_2 \cup H_3)$ with blue edges to H_1 and H_3 and red edges to H_2, let B be the set with red edges to H_2 and H_3 and blue edges to H_1, and let C be the set with blue edges to H_1 and H_2 and red edges to H_3. Note that any or all of these sets of vertices may be empty and $G = H_1 \cup H_2 \cup H_3 \cup A \cup B \cup C \cup T$. Also note that $T_1 = T_2 = \emptyset$.

Either A or C must be empty since the blue edges between H_2 and H_3 along with a blue path of the form $H_2 C H_1 A H_3$ yields a blue copy of $C_{2\ell+1}$. Without loss of generality, suppose $C = \emptyset$. Each part of H_2 and H_3 contains no red or blue edges and the set $H_1 \cup A \cup B$ contains no blue edges. Let $H_2' = H_2 \cup T_3$ and let $H_3' = H_3 \cup T_4 \cup T_5 \cup \cdots \cup T_k$ so by Lemma 3.4 (which is made possible by the fact that $|T_i| \le 2\ell$) and induction on k, we have $|H_i'| \le gr_{k-2}(K_3 : C_{2\ell+1}) - 1$ for $i \in \{2, 3\}$. This yields

$$
\begin{aligned}
n &= |H_2'| + |H_3'| + |H_1 \cup A \cup B| \\
&\le 2[gr_{k-2}(K_3 : C_{2\ell+1}) - 1] + [gr_{k-1}(K_3 : C_{2\ell+1}) - 1] \\
&= 2(\ell \cdot 2^{k-2}) + (\ell \cdot 2^{k-1}) \\
&< \ell \cdot 2^k + 1 \\
&\le gr_k(K_3 : C_{2\ell+1}),
\end{aligned}
$$

a contradiction, completing the proof of this case.

Case 3.4 $r = 2$.

Without loss of generality, suppose the edges between H_1 and H_2 are red. Let A be the set of vertices in $(G \setminus T) \setminus (H_1 \cup H_2)$ with red edges to H_1 and blue edges to H_2, let B be the set with blue edges to $H_1 \cup H_2$, and let C be the set with blue edges to H_1 and red edges to H_2. No vertex $b \in B$ can have red edges to both $a \in A$ and $c \in C$ since the red edges between H_1 and H_2 along with a red path of the form $H_1 abc H_2$ would yield a red copy of $C_{2\ell+1}$ in G. Note that $T_1 = \emptyset$. If $|A \cup B| \ge \ell + 1$ (or similarly $|B \cup C| \ge \ell + 1$), then neither $A \cup B$ nor H_2 (respectively, neither $B \cup C$ nor H_1) can contain any blue edges.

If both $|A \cup B| \ge \ell + 1$ and $|B \cup C| \ge \ell + 1$, then either $B = \emptyset$ or $|B| > \ell$ and $\{|A|, |C|\} = \{0, 1\}$. Assuming the first case, with $|A| \ge \ell + 1$ and $|C| \ge \ell + 1$, A and C must also contain no red edges. With no red or blue edges, A and C must each be single parts of the Gallai partition, contradicting the assumption that $r = 2$ (since $|A|, |C| \ge \ell + 1$). In the latter case, we have $T_1 = T_2 = \emptyset$ and the calculations are again easy.

Next suppose one of $|A \cup B|$ or $|B \cup C|$ is at least $\ell + 1$ and the other is not, say $|A \cup B| \ge \ell + 1$. Suppose further that $|A| \ge \ell + 1$ so A contains no red or blue edges. By Lemma 3.4 (which is made possible by the fact that $|T_i| \le 2\ell$), we get

$$
\begin{aligned}
n &= |H_1 \cup B \cup C \cup T_2| + |H_2 \cup T_k| + |A \cup T_3 \cup \cdots \cup T_{k-1}| \\
&\leq [gr_{k-1}(K_3 : C_{2\ell+1}) - 1] + 2[gr_{k-2}(K_3 : C_{2\ell+1}) - 1] \\
&= (\ell \cdot 2^{k-1}) + 2(\ell \cdot 2^{k-2}) \\
&< \ell \cdot 2^k + 1 \\
&\leq gr_k(K_3 : C_{2\ell+1}),
\end{aligned}
$$

a contradiction, meaning that we may assume $|A| \leq \ell$. Then again using Lemma 3.4, we get

$$
\begin{aligned}
n &= |H_1 \cup B \cup C \cup T_k| + |H_2 \cup A \cup T_2 \cup T_3 \cup \cdots \cup T_{k-1}| \\
&\leq 2[gr_{k-1}(K_3 : C_{2\ell+1}) - 1] \\
&= 2(\ell \cdot 2^{k-1}) \\
&< \ell \cdot 2^k + 1 \\
&\leq gr_k(K_3 : C_{2\ell+1}),
\end{aligned}
$$

a contradiction. In fact, the same analysis as this last subcase also applies when both $|A \cup B| \leq \ell$ and $|B \cup C| \leq \ell$, completing the proof of this case.

Case 3.5 $r = 1$.

Let A be the set of vertices in $(G \setminus T) \setminus H_1$ with red edges to H_1 and let B be the set with blue edges to H_1. If $|A| \leq \ell$ and $|B| \leq \ell$, we move both A and B to T, contradicting the maximality of T. If one of these sets is large and the other is not, say $|A| \geq \ell + 1$ and $|B| \leq \ell$, then neither H_1 nor A can contain any red edge (and $T_1 = \emptyset$). By induction on k, this means that $|A| \leq gr_{k-1}(K_3 : C_{2\ell+1}) - 1$ and so we may apply Lemma 3.4 (which is made possible by the fact that $|T_i| \leq 2\ell$) to get $|A \cup T_2 \cup T_3 \cup \cdots \cup T_{k-1}| \leq gr_{k-1}(K_3 : C_{2\ell+1}) - 1$. Additionally, by Lemma 3.4, we have $|H_1 \cup B \cup T_k| \leq gr_{k-1}(K_3 : C_{2\ell+1}) - 1$. Putting these together, we get

$$
\begin{aligned}
n &= |A \cup T_2 \cup T_3 \cup \cdots \cup T_{k-1}| + |H_1 \cup B \cup T_k| \\
&\leq 2[gr_{k-1}(K_3 : C_{2\ell+1}) - 1] \\
&= 2(\ell \cdot 2^{k-1}) \\
&< \ell \cdot 2^k + 1 \\
&\leq gr_k(K_3 : C_{2\ell+1}),
\end{aligned}
$$

a contradiction.

Finally suppose $|A| \geq \ell + 1$ and $|B| \geq \ell + 1$ so H_1 contains neither red nor blue edges, A contains no red edge, B contains no blue edge, and $T_1 = T_2 = \emptyset$. Note that A and B consist of parts of the Gallai partition with order at most ℓ. By Lemma 3.5, we know that $|A| \leq 2\ell$ and $|B| \leq 2\ell$. By Lemma 3.4, we have $|H_1 \cup T_3 \cup T_4 \cup \cdots \cup T_{k-1}| \leq gr_{k-2}(K_3 : C_{2\ell+1}) - 1$ and $|A \cup T_k| \leq 3\ell$. Putting these together, we get

$$n = |H_1 \cup T_3 \cup T_4 \cup \cdots \cup T_{k-1}| + |A \cup T_k| + |B|$$
$$\le [gr_{k-2}(K_3 : C_{2\ell+1}) - 1] + 3\ell + 2\ell$$
$$= (\ell \cdot 2^{k-2}) + 5\ell$$
$$< \ell \cdot 2^k + 1$$
$$\le gr_k(K_3 : C_{2\ell+1}),$$

a contradiction. This completes the proof of Case 3.5 and Claim 3.3.1.   □

By Claim 3.3.1, we know that all parts in the Gallai partition of $G \setminus T$ have at most ℓ vertices. Since no such part can contain a copy of $C_{2\ell+1}$ this means we may assume $k = 3$ and so $n = \ell \cdot 2^3 + 1 = 8\ell + 1$.

Claim $|T_3| \le \frac{\ell}{2}$.

Proof Suppose, for a contradiction, that $|T_3| > \frac{\ell}{2}$ and call color 3 green. Let $m = |T_3|$ so by Claim 3.3.1, this means that $\frac{\ell}{2} < m \le \ell$. There can be at most $2\ell - 3m$ disjoint copies of P_3 in green in $G \setminus T$ since otherwise the vertices of T_3 could be used (since all edges from T_3 to $G \setminus T$ are green) along with these copies of P_3 to create a green copy of $C_{2\ell+1}$. This means we can delete a set F of at most $6\ell - 9m$ vertices from $G \setminus T$ to leave behind at most a matching in green. This means that there is a Gallai partition of $G \setminus (T \cup F)$ in which all parts have order at most 2. We consider such a partition for the remainder of this proof. Let $m_i = |T_i|$ for $i \in \{1, 2\}$ so since $m > \frac{\ell}{2}$, we get

$$|G \setminus (T \cup F)| \ge (8\ell + 1) - (6\ell - 9m) - m - m_1 - m_2$$
$$> (6\ell + 1) - m_1 - m_2.$$

By Claim 3.3.1, we must have $m_i = 0$ for some i, so suppose $m_1 > 0$ and $m_2 = 0$, and so we wish to find either a red copy of $C_{2(\ell-m_1+1)}$ or a blue copy of $C_{2\ell+1}$ in $G \setminus (T \cup F)$. There are $(6\ell + 1) - m_1$ vertices remaining in $G \setminus (T \cup F)$. By Theorem 3.17, we have

$$R(C_{2\ell+1}, C_{2(\ell-m_1+1)}) = \max\{3\ell - m_1 + 1, 4\ell - 4m_1 + 3\}.$$

First consider the subcase where $3\ell - m_1 + 1 > 4\ell - 4m_1 + 3$, so this means there are at most a total of $3\ell - m_1$ parts in the Gallai partition of $G \setminus (T \cup F)$ to avoid the desired monochromatic cycle. With at least $6\ell + 1 - m_1$ vertices but all parts having order at most 2, this means there are at least

$$[6\ell - 1 - m_1] - [3\ell - m_1] = 3\ell - 1$$

parts of order 2, so there are at least $6\ell - 2$ vertices in parts of order 2. With only $6\ell - 1 - m_1$ vertices available, this implies that $m_1 = 1$. Conversely, the assumption that $3\ell - m_1 + 1 > 4\ell - 4m_1 + 3$ with $m_1 = 1$ yields $3\ell > 4\ell - 1$, a contradiction since $\ell \ge 3$. We may therefore assume that $3\ell - m_1 + 1 \le 4\ell - 4m_1 + 3$. This means there are at most $4\ell - 4m_1 + 2$ parts in the Gallai partition of $G \setminus (T \cup F)$. We must therefore have at least

$$[6\ell - 1 - m_1] - [4\ell - 4m_1 + 2] = 2\ell + 3m_1 - 1$$

parts of order 2 in the partition. Within the reduced graph restricted to the parts of order 2, it suffices to find either a blue odd cycle $C_{\ell+1}$ or $C_{\ell+2}$ (whichever is odd) or a red even cycle $C_{\ell-m_1+1}$ or $C_{\ell-m_1+2}$ (whichever is even). With the given assumptions, Theorem 3.17 implies that there can be at most $2\ell - 2m_1 + 2$ parts of order 2. Putting these last observations together, we arrive at $5m_1 \leq 3$, a contradiction.

Finally suppose $m_1 = m_2 = 0$. Then $|G \setminus (T \cup F)| \geq 6\ell + 2$. Let r be the number of parts of order 2 in the Gallai partition of $G \setminus (T \cup F)$. By Theorem 3.17, we know that $r \leq 2\ell + 2$ since otherwise there would exist a monochromatic odd cycle of length $\ell + 1$ or $\ell + 2$ (whichever is odd) in the reduced graph, making a monochromatic copy of $C_{2\ell+1}$ in G. Since there are at least $6\ell + 2$ vertices, this means there are actually at least

$$(6\ell + 2) - 2(2\ell + 2) = 2\ell - 2$$

parts of order 1. Since $\ell \geq 3$, there is at least one such part, say X. If $r = 2\ell + 2$, then the reduced graph on these parts along with x must produce a monochromatic odd cycle of length $\ell + 1$ or $\ell + 2$, making a monochromatic copy of $C_{2\ell+1}$ in G. This means $r \leq 2\ell + 1$ so there are at least

$$(6\ell + 2) - 2(2\ell + 1) = 2\ell$$

parts of order 1. Therefore, there must be at least $(2\ell + 1) + 2\ell = 4\ell + 1$ parts in the Gallai partition. By Theorem 3.17, there is a monochromatic copy of $C_{2\ell+1}$ in the reduced graph, and therefore in G. $\square$

Our final claim shows that the largest part H_1 is even smaller than previously claimed.

Claim $|H_1| \leq \frac{\ell}{2}$.

Proof Let G_R (and G_B) be the sets of vertices in $(G \setminus T) \setminus (H_1 \cup T_3)$ with red (or blue, respectively) edges to H_1, say with $|G_R| \geq |G_B|$. We first show that $s = |H_1| \leq \frac{2\ell}{3}$ so suppose that $\frac{2\ell+1}{3} \leq s \leq \ell$. By Claim 3.3.1, we know that $|T_3| \leq \frac{\ell}{2}$ so

$$G_R \geq \frac{(8\ell + 1) - \frac{\ell}{2} - s}{2}.$$

If there are at least $2\ell + 1 - 2s$ disjoint red edges within G_R, then there is a red copy of $C_{2\ell+1}$ by using a red complete bipartite graph between H_1 and G_R and

Fig. 3.10 Construction of a red copy of $C_{2\ell+1}$

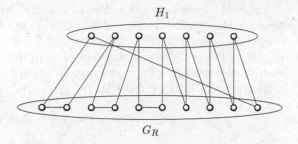

including these disjoint edges. See Fig. 3.10 for an example of this construction. We may therefore delete at most $2(2\ell - 2s)$ vertices from G_R to produce a subgraph G'_R containing no red edges with

$$|G'_R| \geq \frac{8\ell + 1 - \frac{\ell}{2} - s}{2} - 4\ell + 4s$$

$$= \frac{14s - \ell + 2}{4}$$

$$\geq \frac{25\ell + 20}{12} > 2\ell + 1$$

so by Lemma 3.5, there is a blue copy of $C_{2\ell+1}$ within G'_R, a contradiction. We may therefore assume that $\frac{\ell+1}{2} \leq |H_1| \leq \frac{2\ell}{3}$.

Next we show that $|G_B| \geq 2\ell$. Otherwise $|G_B| \leq 2\ell - 1$ so $|G_R| \geq 8\ell + 1 - \frac{\ell}{2} - s - (2\ell - 1)$. Let P^1 be a longest red path within G_R so $|P^1| < 2(\ell - s) + 1$ to avoid creating a red copy of $C_{2\ell+1}$ using a construction similar to the above, much like Fig. 3.10. Note that $|P^1| \geq 2$ since there is at least one red edge in G_R (by Lemma 3.5) and P_1 is a longest red path. Since P^1 is a longest red path, its end vertices u and v, say with $u \in H_u$ and $v \in H_v$, must have no red edges to $F = G_R \setminus (P^1 \cup H_u \cup H_v)$ (recall that $T_1 = T_2 = \emptyset$). Then

$$|F| \geq 8\ell + 1 - \frac{\ell}{2} - s - (2\ell - 1) - |P^1| - 2(s - 1).$$

This means that for every vertex $x \in F$, say with $x \in H_x$, the degree of x within F in red or blue is at least

$$|F| - |H_x| \geq \left[8\ell + 1 - \frac{\ell}{2} - s - (2\ell - 1) - |P^1| - 2(s - 1) \right] - s$$

$$\geq [2(\ell - s) + 3 - |P^1|] + [2\ell - 2] - 3.$$

By Lemma 3.6, there is either a red copy of $P_{2(\ell-s)+3-|P^1|}$ or a blue copy of $P_{2\ell-2}$ within F, either case resulting in a monochromatic copy of $C_{2\ell+1}$. This means we may assume $|G_B| \geq 2\ell$.

Let P^1 be a longest red path within G_R and let P^2 be a longest blue path within G_B. Note that $|P^i| \leq 2(\ell - s) + 1 \leq \ell$ for $i \in \{1, 2\}$ to avoid creating a monochromatic copy of $C_{2\ell+1}$ and $|P^1|, |P^2| \geq 2$ by Lemma 3.5. Let $v_1 \in T_3$ (if $T_3 \neq \emptyset$), $v_2, v_3 \in H_1$, $v_4 \in G_R \setminus P^1$, $v_5 \in G_B \setminus P^2$, and $v_6 v_7$ be any green edge within $G \setminus T_3$ (note that such an edge must exist since otherwise $G \setminus T_3$ is 2-colored with more than $4\ell + 1$ vertices). Let $G' = G \setminus (P^1 \cup P^2 \cup \{v_1, v_2, \ldots, v_7\})$ so

$$|G'| \geq (8\ell + 1) - (2\ell + 7) > 6\ell - 9.$$

By Theorem 3.18, there is a monochromatic copy of $C_{2\ell-2}$ in G'. Regardless of the color (red, blue, or green) and location with respect to the sets (i.e. H_1, G_R, G_B, ...), we may construct a monochromatic copy of $C_{2\ell+1}$ using this monochromatic copy of $C_{2\ell-2}$ and some vertices of $G \setminus G'$, a contradiction. □

Let G_R and G_B denote the sets of vertices in $G \setminus H_1$ with red and blue edges, respectively, to H_1. Without loss of generality, suppose $|G_R| \geq |G_B|$.

Next we show that $|G_B| \geq \frac{5\ell}{2} + 6$. Recall that $s = |H_1|$. Otherwise $|G_B| \leq \frac{5\ell}{2} + 5$ so $|G_R| \geq 8\ell + 1 - \frac{\ell}{2} - s - \left(\frac{5\ell}{2} + 5\right)$. Let P^1 be a longest red path within G_R so $|P^1| < 2(\ell - s) + 1$ to avoid creating a red copy of $C_{2\ell+1}$. Note that $|P^1| \geq 2$ by Lemma 3.5. Since P^1 is a longest red path, its end vertices u and v, say with $u \in H_u$ and $v \in H_v$, must have no red edges to $F = G_R \setminus (P^1 \cup H_u \cup H_v)$. Then

$$|F| \geq 8\ell + 1 - \frac{\ell}{2} - s - \left(\frac{5\ell}{2} + 5\right) - |P^1| - 2(s - 1).$$

For every vertex $x \in F$, the degree of x within F restricted to red and blue is at least

$$8\ell + 1 - \frac{\ell}{2} - s - \left(\frac{5\ell}{2} + 5\right) - |P^1| - 2(s - 1) - s \geq [2(\ell - s) + 3 - |P^1|] + [2\ell - 2] - 3.$$

By Lemma 3.6, there is either a red copy of $P_{2(\ell-s)+3-|P^1|}$ or a blue copy of $P_{2\ell-2}$ within F, either case resulting in a monochromatic copy of $C_{2\ell+1}$. This means we may assume $|G_B| \geq 5\ell/2 + 6$.

Let P^1 be a longest red path within G_R and let P^2 be a longest blue path within G_B. Note that $|P^i| \leq 2(\ell - s) + 1$ for $i \in \{1, 2\}$ to avoid creating a monochromatic copy of $C_{2\ell+1}$ and $|P^1|, |P^2| \geq 2$ by Lemma 3.5. Let H_{R_1} and H_{R_2} be the parts of the Gallai partition of $G \setminus T$ which contain the end vertices of P^1 (where it is possible that $H_{R_1} = H_{R_2}$) and similarly let H_{B_1} and H_{B_2} be the parts containing the end vertices of P^2. Let $u_1 u_2 \ldots u_{|P^1|}$ be the vertices of P^1 and let $v_1 v_2 \ldots v_{|P^2|}$ be the vertices of P^2.

Let $G'_B = G_B \setminus (P^2 \cup H_{B_1} \cup H_{B_2})$ so $|G'_B| \geq (\frac{5\ell}{2} + 6) - |P^2| - 2s + 2 = [2(\ell - s) + 1 - |P^2|] + [\frac{\ell}{2} + 10] - 3$. By Lemma 3.6, there is either a blue path of order $2(\ell - s) + 1 - |P^2|$ or a red path of order at least $\frac{\ell}{2} + 10$. The blue path, in combination with P^2 and H_1, would produce a blue copy of $C_{2\ell+1}$ so we may assume there is a red path in G_B of order at least $\frac{\ell}{2} + 10$. By the same argument, there is a blue path of order at least $\frac{\ell}{2} + 10$ in G_R since $|G_R| \geq |G_B|$. In particular, these observations mean that there is a red edge $x_1 y_1$ in $G_B \setminus (P^2 \cup H_{B_1} \cup H_{B_2})$ and a blue edge $x_2 y_2$ in $G_R \setminus (P^1 \cup H_{R_1} \cup H_{R_2})$.

Let $w_1 \in T_3$ (if $T_3 \neq \emptyset$), $w_2 w_3$ be any green edge within $G \setminus T_3$ (note that such an edge must exist since otherwise $G \setminus T_3$ is 2-colored with more than $4\ell + 1$ vertices), $w_4, w_5 \in H_1$, and let $x_1 y_1$ and $x_2 y_2$ be defined as above. Let $Q^1 = \{u_1, u_2, \ldots, u_{\ell-1}, u_{|P^1|}\}$ and let $Q^2 = \{v_1, v_2, \ldots, v_{\ell-1}, v_{|P^2|}\}$ where some vertices in these sets may not exist if $|P^i| < \ell$. Finally let

$$G' = G \setminus (\{w_1, w_2, w_3, w_4, w_5, x_1, y_1, x_2, y_2\} \cup Q^1 \cup Q^2)$$

so $|G'| \geq |G| - (9 + 2\ell) > 6\ell - 9$. By Theorem 3.18, there exists a monochromatic copy of $C_{2\ell-2}$ within G', say C.

If C is green, then since all parts of the Gallai partition of $G \setminus T$ have order at most $\frac{\ell}{2}$, C must use edges from T_3 to $G' \setminus T_3$. Let $e = uv$ be one such edge with $u \in T_3$ and $v \notin T_3$. Then replacing the edge uv with the path $u w_2 w_3 w_1 v$ produces a green copy of $C_{2\ell+1}$.

Since we are no longer applying the assumption that $|G_R| \geq |G_B|$, we may assume, without loss of generality, that C is red and claim that a symmetric argument would hold if C was blue. If C contains two vertices in G_R, say u and v, at distance (along C) at most 2, then replacing this path with the red path $u w_4 u_1 w_5 v$ or $u w_4 u_1 u_2 w_5 v$ (of the appropriate length) produces a red copy of $C_{2\ell+1}$. Thus, there can be no two vertices in $C \cap G_R$ at distance at most 2 along C. More generally, if P^1 is longer then there can be no two vertices u and v in $G_R \cap C$ at distance at most $|P^1|$ along C.

Similarly, if there is an edge of C from H_1 to G_R, say uv with $u \in H_1$ and $v \in G_R$, then we replace this edge with the red path $u u_1 u_2 w_4 v$ to produce a red copy of $C_{2\ell+1}$. This means there is no edge from H_1 to G_R on C and therefore, $C \cap H_1 = \emptyset$.

If C contains two vertices u and v with $u, v \in G_B \setminus (P^2 \cup H_{B_1} \cup H_{B_2})$ at distance at most 2 on C, then replacing the edge with a red path $u v_1 x_1 v_{|P^2|} v$ or $u v_1 x_1 y_1 v_{|P^2|} v$ of the appropriate length produces a red copy of $C_{2\ell+1}$. This means C cannot have two vertices at distance at most 2 in $G_B \setminus (P^2 \cup H_{B_1} \cup H_{B_2})$. If C contains any vertex $u \in H_{B_i}$ for $i \in \{1, 2\}$, then since C cannot be contained entirely within H_{B_i}, there is an edge of C from H_{B_i} to $G \setminus H_{B_i}$, say uv where $v \notin H_{B_i}$. Supposing $i = 1$ without loss of generality, we can replace the edge uv with the red path $u x_1 y_1 v_1 v$ to produce a red copy of $C_{2\ell+1}$. We therefore know that $C \cap (H_{B_1} \cup H_{B_2}) = \emptyset$.

If $|P^1| \geq 5$, then since no two vertices at distance at most 5 along C can be in G_R and no two vertices at distance at most 2 along C can be in $G_B \setminus (P^2 \cup H_{B_1} \cup H_{B_2})$, at most $\frac{|C|}{6} + \frac{|C|}{3}$ of the vertices of C can be in $G' \setminus P^2$. This leaves at least $\frac{5|C|}{10} = \frac{|C|}{2}$ of the vertices of C contained in P^2. Since $|P^2 \cap G'| \leq 2(\ell - s) + 1 - \ell$ and $|C| = 2\ell - 2$, this is impossible. Thus, we may assume $|P^1| \leq 4$.

At this point, we may redefine Q^2 to be $Q^{2*} = \{v_1, v_2, \ldots, v_{2\ell-5}, v_{|P^2|}\}$ and redefine G' to be

$$G'^* = G \setminus (\{w_1, w_2, w_3, w_4, w_5, x_1, y_1, x_2, y_2\} \cup P^1 \cup Q^{2*})$$

and apply the same arguments as above to arrive at a contradiction. This completes the proof of Theorem 3.14. □

3.3.2   Complete Graphs

Fox et al. conjectured the precise value of the Gallai-Ramsey number for a rainbow triangle or a monochromatic complete graph.

Conjecture 1 ([26]) For $k \geq 1$ and $p \geq 3$,

$$gr_k(K_3 : K_p) = \begin{cases} (r(p) - 1)^{k/2} + 1 & \text{if } k \text{ is even,} \\ (p - 1)(r(p) - 1)^{(k-1)/2} + 1 & \text{if } k \text{ is odd.} \end{cases}$$

Theorem 3.12 provides the solution for the triangle. For the next case, the following was shown by Liu et al. for a monochromatic K_4.

Theorem 3.19 ([27]) For $k \geq 1$,

$$gr_k(K_3 : K_4) = \begin{cases} 17^{k/2} + 1 & \text{if } k \text{ is even,} \\ 3 \cdot 17^{(k-1)/2} + 1 & \text{if } k \text{ is odd.} \end{cases}$$

In order to prove Theorem 3.19, the authors actually prove the following refined version. Theorem 3.19 follows as a corollary to Theorem 3.20 by choosing $s = k$. For this statement, recall that $gr_k(K_3 : sH_1, (k - s)H_2)$ is the minimum integer n such that every k-coloring of the edges of K_n contains either a rainbow triangle, a monochromatic copy of H_1 in one of the first s colors or a monochromatic copy of H_2 in one of the remaining $k - s$ colors.

Theorem 3.20 ([27]) Let $k \geq 1$, and s be an integer with $0 \leq s \leq k$. Then

$$gr_k(K_3 : sK_4, (k - s)K_3) = g(k, s)$$

where

$$g(k,s) = \begin{cases} 17^{s/2} \cdot 5^{(k-s)/2} + 1 & \textit{if } s \textit{ and } (k-s) \textit{ are both even,} \\ 2 \cdot 17^{s/2} \cdot 5^{(k-s-1)/2} + 1 & \textit{if } s \textit{ is even and } (k-s) \textit{ is odd,} \\ 3 \cdot 17^{(k-1)/2} + 1 & \textit{if } s = k \textit{ and } s \textit{ is odd,} \\ 8 \cdot 17^{(s-1)/2} \cdot 5^{(k-s-1)/2} + 1 & \textit{if } s \textit{ and } (k-s) \textit{ are both odd,} \\ 16 \cdot 17^{(s-1)/2} \cdot 5^{(k-s-2)/2} + 1 & \textit{if } s < k \, s \textit{ is odd, and } (k-s) \textit{ is even.} \end{cases}$$

We include the proof of the lower bound by Liu et al. [27] since the proof of the upper bound is long and technical.

Proof (*lower bound*) We first prove the lower bound of Theorem 3.20 by construction. For this construction, we will use the sharpness examples from classical Ramsey results. For $i, j \in \{3, 4\}$, let $H_{i,j}$ be a sharpness example of order $r(i, j) - 1$. In particular, $|H_{4,4}| = 17$, $|H_{4,3}| = 8$, and $|H_{3,3}| = 5$. We construct our sharpness example by taking *blow-ups* of these graphs, that is, replacing each vertex with a particular graph and correspondingly replacing each edge with a complete bipartite graph in the same color.

Let G_0 be a single vertex and we iteratively construct Gallai colored graphs G_i using i colors forbidding the appropriate monochromatic subgraphs for the first i colors in the statement. For induction, suppose we have constructed G_i. If $i = k$, then the construction is completed. Otherwise, we consider the following cases.

- If $i \leq s - 2$, construct G_{i+2} by making 17 copies of G_i and inserting each in place of a vertex in a blow-up of $H_{4,4}$ using colors $i + 1$ and $i + 2$.
- If $i = s - 1$ and $k = s$, then construct G_{i+1} by making 3 copies of G_i and inserting each in place of a vertex in a blow-up of K_3 using color $i + 1$.
- If $i = s - 1$ and $k > s$, then construct G_{i+2} by making 8 copies of G_i and inserting each in place of a vertex in a blow-up of $H_{4,3}$ using colors $i + 1$ and $i + 2$.
- If $i \geq s$ and $i = k - 1$, then construct G_{i+1} by making 2 copies of G_i and inserting each in place of a vertex in a blow-up of K_2 using color $i + 1$.
- If $i \geq s$ and $i \leq k - 2$, then construct G_{i+2} by making 5 copies of G_i and inserting each in place of a vertex in a blow-up of $H_{3,3}$ using colors $i + 1$ and $i + 2$.

By construction, it is clear that G_k is a Gallai coloring and it contains no copy of K_4 in any of the first s colors and no copy of K_3 in any of the remaining $k - s$ colors. The order of G_k then follows the theorem statement as

$$|G_k| = \begin{cases} g_1(k,s) & \textit{if } s \textit{ and } (k-s) \textit{ are both even,} \\ g_2(k,s) & \textit{if } s \textit{ is even and } (k-s) \textit{ is odd,} \\ g_3(k,s) & \textit{if } s = k \textit{ and } s \textit{ is odd,} \\ g_4(k,s) & \textit{if } s \textit{ and } (k-s) \textit{ are both odd,} \\ g_5(k,s) & \textit{if } s < k, \textit{ and } s \textit{ is odd, and } (k-s) \textit{ is even.} \end{cases}$$

For the remainder of the proof, we refer the interested reader to [27]. □

The case of K_5 has also recently been settled by Magnant and Schiermeyer. For this statement, let $R = R_2(K_5) - 1$.

Theorem 3.21 ([28]) *For any integer $k \geq 2$,*

$$gr_k(K_3 : K_5) = \begin{cases} R^{k/2} + 1 & \text{if } k \text{ is even,} \\ 4 \cdot R^{(k-1)/2} + 1 & \text{if } k \text{ is odd} \end{cases}$$

unless $R = 42$, in which case we have

$$\begin{cases} gr_k(K_3 : K_5) = 43 & \text{if } k = 2, \\ 42^{k/2} + 1 \leq gr_k(K_3 : K_5) \leq 43^{k/2} + 1 & \text{if } k \geq 4 \text{ is even,} \\ 169 \cdot 42^{(k-3)/2} + 1 \leq gr_k(K_3 : K_5) \leq 4 \cdot 43^{(k-1)/2} + 1 & \text{if } k \geq 3 \text{ is odd.} \end{cases}$$

Note that if $R = 43$, then Theorem 3.21 implies that Conjecture 1 is false. Also recall the following well-known conjecture by McKay and Radziszowski about the sharp value for the 2-color Ramsey number of K_5.

Conjecture 2 ([29]) $R(K_5, K_5) = 43$.

By Theorem 3.21, it turns out that exactly one of Conjecture 1 or Conjecture 2 must be false and the other must be true. This is demonstrated in the following example implying that if Conjecture 2 is true, then Conjecture 1 is not true.

Lemma 3.7 ([28]) *There exists a 3-colored copy of K_{169} which contains no rainbow triangle and no monochromatic copy of K_5.*

Proof Let G_{rb} be a sharpness example on 13 vertices for the Ramsey number $R(K_3, K_5) = 14$ say using colors red and blue, respectively. Such an example as G_{rb} is 4-regular in red and 8-regular in blue. Similarly, let G_{rg} be a copy of the same graph with all blue edges replaced by green edges. We construct the desired graph G by making 13 copies of each vertex in G_{rb} and, for each set of copies (corresponding to a vertex), insert a copy of G_{rg}. If an edge uv in G_{rb} is red (respectively, blue), then all edges in G between the two inserted copies of G_{rg} corresponding to u and v are colored red (respectively, blue). Then G contains no rainbow triangle by construction but also contains no monochromatic K_5. Since $|G| = 169$, this provides the desired example. □

In particular, note that if $R(K_5, K_5) = 43$ so $R = 42$, then Conjecture 1 claims that $gr_3(K_3 : K_5) = 169$ but this example refutes this claim.

Much like the use of Theorem 3.20 to prove Theorem 3.19, in order to prove Theorem 3.21, the authors actually prove the following more refined version, stated in Theorem 3.22. Note that Theorem 3.21 follows from Theorem 3.22 by setting $r = k$, $s = 0$ and $t = 0$.

Theorem 3.22 ([28]) *For nonnegative integers r, s, t, let $k = r + s + t$. Then*

$$gr_k(K_3 : rK_5, sK_4, tK_3)$$

$$= \begin{cases} R^{r/2} \cdot 17^{s/2} \cdot 5^{t/2} + 1 & if\, r, s, t\ are\ even, \\ 2 \cdot R^{r/2} \cdot 17^{s/2} \cdot 5^{(t-1)/2} + 1 & if\, r, s\ are\ even,\ and\ t\ is\ odd, \\ 3 \cdot R^{r/2} \cdot 17^{(s-1)/2} + 1 & if\, r\ is\ even,\ s\ is\ odd,\ and\ t = 0, \\ 4 \cdot R^{(r-1)/2} + 1 & if\, r\ is\ odd,\ and\ s = t = 0, \\ 8 \cdot R^{r/2} \cdot 17^{(s-1)/2} \cdot 5^{(t-1)/2} + 1 & if\, r\ is\ even,\ and\ s, t\ are\ odd, \\ 13 \cdot R^{(r-1)/2} \cdot 17^{s/2} \cdot 5^{(t-1)/2} + 1 & if\, r, t\ are\ odd,\ and\ s\ is\ even, \\ 16 \cdot R^{r/2} \cdot 17^{(s-1)/2} \cdot 5^{(t-2)/2} + 1 & if\, r, t\ are\ even,\ t \geq 2,\ and\ s\ is\ odd, \\ 24 \cdot R^{(r-1)/2} \cdot 17^{(s-1)/2} \cdot 5^{t/2} + 1 & if\, r, s\ are\ odd,\ and\ t\ is\ even, \\ 26 \cdot R^{(r-1)/2} \cdot 17^{s/2} \cdot 5^{(t-2)/2} + 1 & if\, r\ is\ odd,\ s\ is\ even,\ t \geq 2\ is\ even, \\ 48 \cdot R^{(r-1)/2} \cdot 17^{(s-1)/2} \cdot 5^{(t-1)/2} + 1 & if\, r, s, t\ are\ odd, \\ 72 \cdot R^{(r-1)/2} \cdot 17^{(s-2)/2} + 1 & if\, r\ is\ odd,\ t = 0,\ and\ s \geq 2\ is\ even. \end{cases}$$

3.3.3 Other Specific Graphs

In this section, we consider some other sharp results for graphs or classes of graphs with precise structures. The first provides a sort of "off diagonal" case of Gallai-Ramsey numbers where we look for either a monochromatic triangle within some set of colors or a monochromatic copy of C_4 within the remaining colors.

Theorem 3.23 ([8]) *Given integers k and s with $k \geq 2$ and $0 \leq s \leq k$, we have*

$$gr_k(K_3 : s \cdot K_3, (k-s) \cdot C_4)$$

$$= \begin{cases} (k-s+3) \cdot 2 \cdot 5^{(s-1)/2} + 1 & if\, s\ is\ odd\ and\ s < k-1, \\ (k-s+3) \cdot 5^{s/2} + 1 & if\, s\ is\ even\ and\ s < k-1, \\ 6 \cdot 5^{(s-1)/2} + 1 & if\, s\ is\ odd\ and\ s = k-1, \\ 3 \cdot 5^{s/2} + 1 & if\, s\ is\ even\ and\ s = k-1, \\ 2 \cdot 5^{(s-1)/2} + 1 & if\, s\ is\ odd\ and\ s = k, \\ 5^{s/2} + 1 & if\, s\ is\ even\ and\ s = k. \end{cases}$$

The following proof is an adaptation of the proof provided by Wu et al. [8].

Proof Note that if we set $s = 0$, the result reduces to Theorem 3.10 and if we set $s = k$, the question reduces to Theorem 3.12.

For simplicity, let K_5^c denote the 2-coloring of K_5 consisting of one copy of C_5 in each color.

In order to prove the lower bound for $k - s > 1$, we first let G_{k-s} be the sharpness example from Theorem 3.10 on $k - s + 3$ vertices using colors $s + 1, s + 2, \ldots, k$. This graph contains no rainbow triangle and, by construction, no monochromatic copy of C_4. Given G_i for some i with $k - s \leq i \leq k - 2$, let G_{i+2} be the graph constructed by taking a G_i-blow-up of K_5^c using two new colors. The graph G_{i+2} certainly contains no rainbow triangle and no monochromatic copy of K_3 in either of the new colors. If s is even, then G_k is the desired construction. If s is odd, we define G_k be the graph constructed by taking a G_{k-1}-blow-up of K_2 in color s. In either case, the graph G_k contains no rainbow triangle, no monochromatic copy of K_3 in any of the first s colors $\{1, 2, \ldots, s\}$, and no monochromatic copy of C_4 in any of the remaining colors $\{s + 1, s + 2, \ldots, k\}$, and

$$|G_k| = \begin{cases} (k - s + 3) \cdot 2 \cdot 5^{(s-1)/2} & \text{if } s \text{ is odd,} \\ (k - s + 3) \cdot 5^{s/2} & \text{if } s \text{ is even.} \end{cases}$$

For the lower bound when $k - s = 1$, we first let G_1' be a monochromatic K_3 in color k. This graph certainly contains no rainbow triangle and it does not even have enough vertices to contain a monochromatic C_4. Starting with this G_1', we use the same construction as above to produce a colored complete graph G_k' containing no rainbow triangle and no monochromatic C_4 with

$$|G_k'| = \begin{cases} 6 \cdot 5^{(s-1)/2} & \text{if } s \text{ is odd,} \\ 3 \cdot 5^{s/2} & \text{if } s \text{ is even.} \end{cases}$$

For the upper bound, let $t = k - s$ be the number of colors in which we would like to find a monochromatic copy of C_4. Let G be a k-coloring of K_n where

$$n = \begin{cases} (t + 3) \cdot 2 \cdot 5^{(s-1)/2} + 1 & \text{if } s \text{ is odd and } s < k - 1, \\ (t + 3) \cdot 5^{s/2} + 1 & \text{if } s \text{ is even and } s < k - 1, \\ 6 \cdot 5^{(s-1)/2} + 1 & \text{if } s \text{ is odd and } s = k - 1, \\ 3 \cdot 5^{s/2} + 1 & \text{if } s \text{ is even and } s = k - 1, \\ 2 \cdot 5^{(s-1)/2} + 1 & \text{if } s \text{ is odd and } s = k, \\ 5^{s/2} + 1 & \text{if } s \text{ is even and } s = k, \end{cases}$$

and suppose, for a contradiction, that G contains no rainbow triangle, no monochromatic K_3 in one of the first s colors, and no monochromatic C_4 in one of the remaining t colors.

Let k' be the number of colors c appearing in $E(G)$ for which the subgraph of G including precisely those edges of color c, call it G_c, has maximum degree $\Delta(G_c) \geq 2$. Note that $k' \leq k$ and if a color does not have $\Delta(G_c) \geq 2$, then certainly there is no monochromatic triangle or C_4 in that color. We may therefore consider $s' \leq s$ and $t' \leq t$ to be the number of colors in which we would like to produce a monochromatic triangle or C_4, respectively. For ease of notation, we call these colors *wasted* since they cannot possibly contain either a monochromatic triangle or a monochromatic copy of C_4. In fact, to avoid a rainbow triangle, these wasted colors together induce subgraph with maximum degree at most 1. The upper bound follows directly from Theorem 3.24 (below) to complete the proof of Theorem 3.23. □

Theorem 3.24 *Given integers* k, k', s' *and* t' *with* $0 \leq k' \leq k$ *and* $s' + t' = k'$, *if*

$$n = n(s', t') = \begin{cases} (t'+3) \cdot 2 \cdot 5^{(s'-1)/2} + 1 & \text{if } s' \text{ is odd and } t' \geq 2, \\ (t'+3) \cdot 5^{s'/2} + 1 & \text{if } s' \text{ is even and } t' \geq 2, \\ 6 \cdot 5^{(s'-1)/2} + 1 & \text{if } s' \text{ is odd and } t' = 1, \\ 3 \cdot 5^{s'/2} + 1 & \text{if } s' \text{ is even and } t' = 1, \\ 2 \cdot 5^{(s'-1)/2} + 1 & \text{if } s' \text{ is odd and } t' = 0, \\ 5^{s'/2} + 1 & \text{if } s' \text{ is even and } t' = 0, \end{cases}$$

then in every Gallai colored complete graph K_N *(where* $N \geq n$*) using* k *colors in which* k' *of the colors have maximum degree at least* 2, *there is either a monochromatic triangle in one of the first* s' *colors* $\{1, 2, \ldots, s'\}$ *or a monochromatic copy of* C_4 *in one of the last* t' *colors* $\{k - t' + 1, k - t' + 2, \ldots, k\}$.

Note that the middle $k - k'$ colors, the set of colors $\{s' + 1, s' + 2, \ldots, k - t'\}$, are wasted.

Proof We prove Theorem 3.24 by induction on k'. Note that as colors are removed from consideration, the colors are always relabeled so that we always seek a monochromatic triangle in one of the first several colors or a monochromatic C_4 in one of the last several colors. More specifically, the induction process considers a subgraph in which the edges of some color are more restricted than in the whole graph. Within the subgraph, the color is no longer able to contain a copy of K_3 or C_4 so the formerly non-wasted color becomes wasted.

Suppose n is as given in the statement of Theorem 3.24. Observe that if $k' = 0$, then every set of three vertices in G induces a rainbow triangle so we may assume $k' \geq 1$.

Consider a Gallai partition of G, say with red and blue being the colors used on edges in between parts in the Gallai partition, and without loss of generality, if only one color appears between the parts, we call it red. Consider such a Gallai partition of G using the smallest number of parts and let $H_1, H_2, \ldots, H_q$ be the parts of this partition where $q \geq 2$. By Lemma 3.1, the case $q = 3$ is impossible. On the other hand, since the reduced graph of this partition is a 2-colored complete graph and by the following fact, we immediately get that $q \leq 6$ and furthermore, for Cases 1 and 2 of the argument below, we get $q \leq 5$.

Fact 3.25 ([23–25, 30–32]) *The Ramsey numbers for combinations of the triangle and the* C_4 *are*

- $R(K_3, K_3) = 6$,

- $R(C_4, C_4) = 6$ and
- $R(K_3, C_4) = 7$.

We consider cases based on where red and blue are in the list of colors, more specifically, whether they appear within the first s' colors or in the last t' colors.

Suppose that at least one of red or blue is a wasted color, say red is wasted. Then since every vertex of the reduced graph has incident edges in both colors, this means no part of the Gallai partition has order at least 2. With at most 7 parts in the partition, this means $n \leq 7$, for a contradiction. Both red and blue must therefore not be wasted colors.

We also assume, for the remainder of the proof, that $t' > 1$. The remaining cases where $0 \leq t' \leq 1$ (or equivalently $k' - 1 \leq s' \leq k'$) follow the same argument (in fact, without needing Case 2) and so are omitted. Note that the induction is unaffected by this omission since substituting 1 for t' in the first two options for the value of $n(s', t')$ yields $8 \cdot 5^{(s'-1)/2} + 1 > 6 \cdot 5^{(s'-1)/2} + 1$ and $4 \cdot 5^{s'/2} + 1 > 3 \cdot 5^{k/2} + 1$, respectively, and similar inequalities hold when using $k' = s'$.

Case 1 *Red and blue are both within the first s' colors.*

By Fact 3.25, this means that $q \leq 5$. First suppose $q = 2$, say with red as the color appearing on all edges from H_1 to H_2. Note that to avoid creating a red triangle, there must be no red edges within H_i for all i. This means that within H_i, since there can no longer by any red edges, s' is reduced by 1 and k' is also reduced by 1. By induction on k' applied within H_i, for $i \in \{1, 2\}$, we have

$$|H_i| \leq n(s' - 1, t') - 1 = \begin{cases} (t' + 3) \cdot 2 \cdot 5^{(s'-2)/2} & \text{if } s' \text{ is even,} \\ (t' + 3) \cdot 5^{(s'-1)/2} & \text{if } s' \text{ is odd.} \end{cases}$$

This means that

$$\begin{aligned} |G| &= |H_1| + |H_2| \\ &\leq 2 \cdot \begin{cases} (t' + 3) \cdot 2 \cdot 5^{(s'-2)/2} & \text{if } s' \text{ is even,} \\ (t' + 3) \cdot 5^{(s'-1)/2} & \text{if } s' \text{ is odd} \end{cases} \\ &= \begin{cases} (t' + 3) \cdot 4 \cdot 5^{(s'-2)/2} & \text{if } s' \text{ is even,} \\ (t' + 3) \cdot 2 \cdot 5^{(s'-1)/2} & \text{if } s' \text{ is odd} \end{cases} \\ &< \begin{cases} (t' + 3) \cdot 5^{s'/2} + 1 & \text{if } s' \text{ is even,} \\ (t' + 3) \cdot 2 \cdot 5^{(s'-1)/2} + 1 & \text{if } s' \text{ is odd} \end{cases} \\ &\leq |G|, \end{aligned}$$

a contradiction.

We may therefore suppose $4 \leq q \leq 5$. By Lemma 3.2, both red and blue induce connected subgraphs of the reduced graph (and of G). Therefore, each vertex of the reduced graph must have at least one incident edge in each color of the partition (red and blue). To avoid creating a monochromatic triangle in red or blue, each part H_i must contain no red or blue edges. By induction on k' applied within H_i for $i \in \{1, 2, \ldots, q\}$, we have

$$|H_i| \leq n(s' - 2, t') = \begin{cases} (t' + 3) \cdot 2 \cdot 5^{(s'-3)/2} & \text{if } s' \text{ is odd,} \\ (t' + 3) \cdot 5^{(s'-2)/2} & \text{if } s' \text{ is even.} \end{cases}$$

This means that

$$\begin{aligned}
|G| &= \sum_{i=1}^{q} |H_i| \\
&\leq 5 \cdot \begin{cases} (t' + 3) \cdot 2 \cdot 5^{(s'-3)/2} & \text{if } s' \text{ is odd,} \\ (t' + 3) \cdot 5^{(s'-2)/2} & \text{if } s' \text{ is even} \end{cases} \\
&< \begin{cases} (t' + 3) \cdot 2 \cdot 5^{(s'-1)/2} + 1 & \text{if } s' \text{ is odd,} \\ (t' + 3) \cdot 5^{s'/2} + 1 & \text{if } s' \text{ is even} \end{cases} \\
&\leq |G|,
\end{aligned}$$

again a contradiction, completing the proof in this case.

Case 2 *Both red and blue are within the last t' colors.*

By Fact 3.25, this means that $q \leq 5$. First suppose $q = 2$, say with red edges from H_1 to H_2. Then one of H_1 or H_2 must have order 1 to avoid making a red C_4, so we assume $|H_2| = 1$. Within H_1, if there is a red P_3, then we get a red C_4 using the red edges to the vertex of H_2. This means that the maximum degree of the red subgraph within H_1 is at most 1. By induction on k' applied within H_1, we get

$$|H_1| \leq n(s', t' - 1) - 1 \leq \begin{cases} (t' + 2) \cdot 2 \cdot 5^{(s'-1)/2} & \text{if } s' \text{ is odd,} \\ (t' + 2) \cdot 5^{s'/2} & \text{if } s' \text{ is even,} \end{cases}$$

where this second inequality is strict if $t' = 2$ (by the definition of $n(s', t')$). This means that

$$\begin{aligned}
|G| &\leq 1 + n(k' - 1, s) - 1 \\
&\leq \begin{cases} (k' - s + 2) \cdot 2 \cdot 5^{(s-1)/2} + 1 & \text{if } s \text{ is odd,} \\ (k' - s + 2) \cdot 5^{s/2} + 1 & \text{if } s \text{ is even} \end{cases} \\
&< |G|,
\end{aligned}$$

a contradiction.

Thus, we suppose $4 \leq q \leq 5$. Again by minimality of q, both red and blue appear on edges incident to every vertex of the reduced graph. To avoid a monochromatic copy of C_4, at most one of the parts, say H_1, may have order at least 2. Then $|H_2| =$

$|H_3| = |H_4| = 1$. Since each of the vertices in these sets H_i (for $2 \leq i \leq 4$) has a single color on all edges to H_1 and only two colors are available (by the structure of the Gallai partition), two of the vertices, say $v \in H_2$ and $w \in H_3$, must have the same color on all edges to H_1. If $|H_1| \geq 2$, the proof is complete since this structure contains a monochromatic copy of C_4. On the other hand, if $|H_1| = 1$, then all parts of the Gallai partition have order at most 1 so $n \leq 5$, again a contradiction.

Case 3 *Exactly one of red or blue is within the first s' colors and the other is within the last t' colors.*

By Fact 3.25, since $R(K_3, C_4) = R(C_4, K_3) = 7$, this means that $q \leq 6$. Without loss of generality, suppose red appears within the first s' colors and blue appears within the last t' colors. Note that both red and blue must appear within the Gallai partition since otherwise this case would reduce to one of the previous cases. In particular, this means that we may assume $4 \leq q \leq 6$. Additionally, Lemma 3.2, both colors must induce a connected subgraph in the reduced graph.

First suppose $q = 4$. Since there is only one 2-coloring of K_4 in which both colors are connected, we may assume that the reduced graph is precisely this 2-coloring of K_4 in which each color induces a P_4. Let H_1, H_2, H_3, H_4 be the parts of the partition corresponding to the vertices of the blue P_4 (in this order) in the reduced graph (see Fig. 3.6 where the darker edges are blue). Then to avoid a blue copy of C_4, we see that $|H_2| = |H_3| = 1$.

Also since all edges between H_1 and H_4 are red, they must each contain no red edges to avoid creating a red triangle. Since H_2 (and H_3) has all blue edges to H_1 (respectively, H_4), the blue subgraph of H_1 (respectively, H_4) must have maximum degree at most 1. By induction on k' applied within H_1 and H_4, we have

$$|H_1|, |H_4| \leq n(s' - 1, t' - 1) - 1.$$

Putting these together, we get

$$\begin{aligned}
|G| &= \sum_{i=1}^{q} |H_i| \\
&\leq 2 + 2 \cdot (n(s' - 1, t' - 1) - 1) \\
&\leq 2 + 2 \cdot \begin{cases} ((t' - 1) + 3) \cdot 2 \cdot 5^{(s'-2)/2} & \text{if } s' \text{ is even,} \\ ((t' - 1) + 3) \cdot 5^{(s'-1)/2} & \text{if } s' \text{ is odd} \end{cases} \\
&= 2 + \begin{cases} (t' + 2) \cdot 4 \cdot 5^{(s'-2)/2} & \text{if } s' \text{ is even,} \\ (t' + 2) \cdot 2 \cdot 5^{(s'-1)/2} & \text{if } s' \text{ is odd} \end{cases} \\
&< |G|,
\end{aligned}$$

a contradiction.

Next suppose $q = 5$, so the reduced graph is a 2-coloring of K_5 containing no red triangle and no blue C_4 but additionally both colors are connected.

First suppose the reduced graph is a red C_5 with complementary blue C_5 (see Fig. 3.5 for example), say with parts $H_1, H_2, \ldots, H_5$ in order around the blue cycle. If any part, say H_2, has order at least 2, then since H_2 has all blue edges to $H_1 \cup H_3$, there is a blue copy of C_4 within this structure. This means that $|H_i| = 1$ for all i, so $n = 5$, a contradiction. Thus, there must be a blue triangle in the reduced graph.

To avoid a blue copy of C_4, each of the remaining two vertices of the reduced graph must have at most one blue edge to the triangle so, to avoid creating a red triangle, the other two vertices must have a blue edge between them (see Fig. 3.11 for an example of this structure where the darker edges are blue). The blue subgraph being connected by assumption, it must contain a copy of P_5 with the addition of an edge from at least one end vertex to the center vertex of the path, to form a triangle.

Let H_1 be the part corresponding to the end vertex of this path not used in the specified triangle. To avoid creating a blue C_4, all sets other than H_1 must have order 1. With at least one vertex outside H_1 having all red edges to H_1, we see that H_1 contains no red edges. Similarly, with at least one vertex outside H_1 having all blue edges to H_1, we see that any blue subgraph of H_1 must have maximum degree at most 1. By induction on k' applied within H_1, we have

$$|H_1| \le n(s' - 1, t' - 1) - 1.$$

Putting this information together, we get

$$
\begin{aligned}
|G| &= \sum_{i=1}^{q} |H_i| \\
&\le 4 + n(s' - 1, t' - 1) - 1 \\
&\le 4 + \begin{cases} ((t' - 1) + 3) \cdot 2 \cdot 5^{(s'-2)/2} & \text{if } s' \text{ is even,} \\ ((t' - 1) + 3) \cdot 5^{(s'-1)/2} & \text{if } s' \text{ is odd} \end{cases} \\
&= 4 + \begin{cases} (t' + 2) \cdot 2 \cdot 5^{(s'-2)/2} + 1 & \text{if } s \text{ is even,} \\ (t' + 2) \cdot 5^{(s'-1)/2} + 1 & \text{if } s' \text{ is odd} \end{cases} \\
&< |G|,
\end{aligned}
$$

a contradiction.

Fig. 3.11 Blue P_5 with an extra edge in the reduced graph

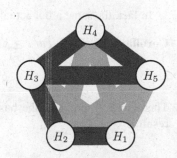

Finally suppose $q = 6$. Since $R(K_3, K_3) = 6$, we know that the reduced graph must contain a monochromatic triangle but we have assumed this is not red so it must be blue. Each remaining vertex of the reduced graph must have at most one blue edge to the blue triangle to avoid creating a blue copy of C_4. This means that all remaining vertices must be adjacent in blue to avoid making a red triangle. Already, the reduced graph contains two disjoint blue triangles which means that, to avoid a blue C_4 in G, all parts of the Gallai partition must have order 1 so $n = 6$, again a contradiction completing the proof.   □

Recall that S_t denotes the star with $t + 1$ total vertices (and therefore t edges). Then for $t \geq 2$, let S_t^+ denote the *star-plus*, the graph consisting of the star S_t with the addition of an edge between two of the pendant vertices, forming a triangle. Note that $S_3^+ = K_3$. Wang et al. [33] found the 2-color Ramsey number for this class of graphs and used that to prove the general Gallai-Ramsey number for the class.

Theorem 3.26 ([33]) *For* $t \geq 3$,

$$R(S_t^+, S_t^+) = 2t + 1.$$

This provides the base for the following Gallai-Ramsey number.

Theorem 3.27 ([33]) *For* $k \geq 1$ *and* $t \geq 2$,

$$gr_k(K_3 : S_t^+) = \begin{cases} 2t \cdot 5^{\frac{k-2}{2}} + 1 & \textit{if } k \textit{ is even,} \\ t \cdot 5^{\frac{k-1}{2}} + 1 & \textit{if } k \textit{ is odd.} \end{cases}$$

Recall that P_t denotes the path of order t. Then for $t \geq 3$, let P_t^+ denote the *path-plus*, the graph consisting of the path P_t with the addition of an edge between one end and the vertex at distance 2 along the path from that end, forming a triangle. Note that $P_3^+ = K_3$ and $P_4^+ = S_4^+$. Wang et al. [33] again found the 2-color Ramsey number for this class of graphs and used that to prove the general Gallai-Ramsey number for the class.

Theorem 3.28 ([33]) *For* $t \geq 4$,

$$R(P_t^+, P_t^+) = 2t - 1.$$

In fact, the same proof actually yields a slightly more general result.

Corollary 3.1 ([33]) *For* $t \geq s \geq 4$,

$$R(P_s^+, P_t^+) = 2t - 1.$$

Theorem 3.28 provides the base for the Gallai-Ramsey number in the following result.

Theorem 3.29 ([33]) *For $t \geq 4$ and $k \geq 1$,*

$$gr_k(K_3 : P_t^+) = \begin{cases} 2(t-1) \cdot 5^{\frac{k-2}{2}} + 1 & \text{if } k \text{ is even,} \\ (t-1) \cdot 5^{\frac{k-1}{2}} + 1 & \text{if } k \text{ is odd.} \end{cases}$$

3.4 General Bounds for Some Classes of Graphs

In this section, we present some general (upper) bounds on the Gallai-Ramsey numbers of some classes of graphs. Naturally, sharpening these bounds to produce precise results would be of great interest.

3.4.1 Even Cycles and Paths

We first consider general bounds on the Gallai-Ramsey number for even cycles by Hall et al.

Theorem 3.30 ([9]) *For all integers k and n with $k \geq 1$ and $n \geq 2$,*

$$(n-1)k + n + 1 \leq gr_k(K_3 : C_{2n}) \leq (n-1)k + 3n.$$

Note that an improvement of the upper bound in this result to $(n-1)k + 2n + 2$ was recently announced by Zhang et al. [20].

We include an adaptation of the proof of Theorem 3.30 by Hall et al. [9]. For this proof, we need two lemmas (Lemmas 3.2 and 3.8).

For two vertex sets A and B, let $A + B$ denote the join of A and B. In other words, $A + B$ is the complete bipartite graph with bipartitions A and B. The operation "$+$" is a binary operation which is not associative. Therefore, let $A_1 + A_2 + \cdots + A_t$ denote the graph $A_1 \cup A_2 \cup \cdots \cup A_t$ with the addition of all edges between A_i and A_{i+1} for all $1 \leq i \leq t - 1$. Let $G_0 := B_1 + A_1 + B_2 + A_2 + B_3$ with $A_1, A_2 \neq \emptyset$, $|B_1| = |B_3|$ and $|B_2| = 1$. See Fig. 3.12, where the heavy edges represent all possible edges between the sets.

Note that G_0 is the bipartite graph consisting of parts $A = A_1 \cup A_2$ and $B = B_1 \cup B_2$.

Lemma 3.8 ([9]) *Let $G = A \cup B$ be a bipartite graph with $|A| \geq 2$, $|B| \geq 4$ and minimum degree of vertices in A satisfying $\delta(A) \geq (|B| + 1)/2$. Let l be an integer with $2 \leq l \leq \min \left\{ |A|, \frac{|B|-1}{2} \right\}$. Then either G contains a copy of C_{2l} or G is isomorphic to G_0.*

For a graph G and an integer l, let $\mathscr{C}_l$ denote the set of cycles in G of length exactly l. The proof of Lemma 3.8 is divided into three cases (see Lemmas 3.9-3.11) depending on the connectivity of G.

Fig. 3.12 Construction of
G_0

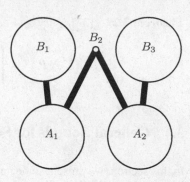

Lemma 3.9 ([9]) *Let $G = A \cup B$ be a connected bipartite graph having a cut vertex in B. If $|A| \geq 2$ and $\delta(A) \geq (|B| + 1)/2$, then $\delta(A) = (|B| + 1)/2$ and G is isomorphic to G_0.*

Proof For this proof, let v be the cut vertex in B and let B_1 be the intersection of one component of $G \setminus v$ with B. Let B_2 be all the remaining vertices of B, those in $B \setminus (B_1 \cup \{v\})$. Since each vertex in A has degree at least $(|B| + 1)/2$ to B, every vertex of A must be adjacent to v and all of either B_1 or B_2 but not both (since v is a cut vertex). This means that G is isomorphic to G_0 and $\delta(A) = (|B| + 1)/2$ as claimed.   □

Lemma 3.10 ([9]) *Let $G = A \cup B$ be a connected bipartite graph having no cut vertex in B. If $|A| \geq 2$, $|B| \geq 4$ and $\delta(A) \geq (|B| + 1)/2$, then $\mathscr{C}_{2l} \neq \emptyset$ for any $2 \leq l \leq \min\{|A|, \delta(A) - 1\}$.*

Before proving Lemma 3.10, we need to define some terminology used in this section. Write a cycle C with a given cyclic orientation by $\overrightarrow{C}$. Let C be a cycle or a path. For $x, y \in V(C)$, we denote by $x \overrightarrow{C} y$ a path from x to y on $\overrightarrow{C}$. The reverse sequence of $x \overrightarrow{C} y$ is denoted by $y \overleftarrow{C} x$. For $x \in V(C)$, we denote the successor and the predecessor of x on $\overrightarrow{C}$ by x^+ and x^-, respectively. For $X \subseteq V(C)$, we define $X^+ := \{x^+ : x \in X\}$ and $X^- := \{x^- : x \in X\}$. For an integer $i \geq 2$, we inductively define $x^{+i} := (x^{+(i-1)})^+$.

Proof of Lemma 3.10. We prove Lemma 3.10 by induction on l. First a claim to provide the base of this induction.

Claim 3.31 $\mathscr{C}_4 \neq \emptyset$.

Proof First suppose that $|A| = 2$, that is, $A := \{a_1, a_2\}$. Since G has no cut vertex in B, $|N_G(a_1) \cap N_G(a_2)| \geq 2$. Then we can see $\mathscr{C}_4 \neq \emptyset$. Hence, we may assume that $|A| \geq 3$, and for a contradiction, suppose that $\mathscr{C}_4 = \emptyset$. Then $|N_G(a_1) \cap N_G(a_2)| \leq 1$ for any $a_1, a_2 \in A$ with $a_1 \neq a_2$. Let $a_1, a_2, a_3 \in A$ be three distinct vertices. Then $|B| \geq |N_G(a_1) \cup N_G(a_2) \cup N_G(a_3)| \geq 3((|B| + 1)/2) - 3$, that is $|B| \leq 3$, a contradiction. Hence, $\mathscr{C}_4 \neq \emptyset$.   □

The next claim provides the induction step of the proof of Lemma 3.10.

Claim 3.32 For any $3 \le l \le \min\{|A|, \delta(A) - 1\}$, if $\mathscr{C}_{2l-2} \neq \emptyset$, then $\mathscr{C}_{2l} \neq \emptyset$.

Proof Suppose, for a contradiction, that $\mathscr{C}_{2l-2} \neq \emptyset$ and $\mathscr{C}_{2l} = \emptyset$ for some l with $3 \le l \le \min\{|A|, \delta(A) - 1\}$. For any cycle $C \in \mathscr{C}_{2l-2}$ and vertex $x \in V(G) - V(C)$, let $H_C := G - C$ and $H_C^x := N_G(x) \cap H_C$. If $A \subseteq V(C)$ for some cycle $C \in \mathscr{C}_{2l-2}$, then $l \le |A| = |C|/2 = l - 1$, a contradiction. Hence, $A \cap H_C \neq \emptyset$ for any cycle $C \in \mathscr{C}_{2l-2}$. Suppose that $|H_C^x| \le 1$ for some cycle $C \in \mathscr{C}_{2l-2}$ and vertex $x \in H_C \cap A$. Then $\delta(A) \le d_G(x) \le |C \cap B| + 1 = l \le \delta(A) - 1$, a contradiction. Hence, $|H_C^x| \ge 2$ for any cycle $C \in \mathscr{C}_{2l-2}$ and vertex $x \in H_C \cap A$.

Suppose that there exist no cycle $C \in \mathscr{C}_{2l-2}$ and $x \in H_C \cap A$ such that $N_C(x) \neq \emptyset$. Let $C \in \mathscr{C}_{2l-2}$. Since G is connected, there exist vertices $x \in H_C \cap A$ and $y \in H_C^x$ such that $N_C(y) \neq \emptyset$. By assumption, $N_C(x) = \emptyset$. For convenience, we abbreviate H_C and H_C^x by H and H^x, respectively. Let $u \in N_C(y)$ and suppose that $y \in N_H(u^{+2})$. Then $C' := u^{+2} \overrightarrow{C} uyu^{+2}$ is a cycle such that $C' \in \mathscr{C}_{2l-2}$, $x \in H_{C'} \cap A$ and $N_{C'}(x) \neq \emptyset$, a contradiction. Hence, we get $y \notin N_H(u^{+2})$. Since $\mathscr{C}_{2l} = \emptyset$, $N_H(u^{+2}) \cap N_H(x) - \{y\} = \emptyset$. Then $|B| = |C \cap B| + |H \cap B| \ge |N_C(u^{+2})| + |N_H(u^{+2})| + |N_H(x)| \ge 2((|B| + 1)/2) = |B| + 1$, a contradiction. Hence, there exists a cycle $C \in \mathscr{C}_{2l-2}$ and a vertex $x \in H_C \cap A$ such that $N_C(x) \neq \emptyset$.

For convenience, we again abbreviate H_C and H_C^x by H and H^x, respectively. Let $v \in N_C(x)$ and $Z := (N_C(v^+)^- \cap N_C(x)^+) - \{v^+\}$. For $z \in Z$, $C' := v^+ z^+ \overrightarrow{C} vxz^- \overleftarrow{C} v^+$ is a cycle such that $C' \in \mathscr{C}_{2l-2}$ and $z \in H_{C'} \cap A$. Hence, for each $z \in Z$, we can take a set $O_z \subseteq H_{C'}^z \subseteq H_C$ with $|O_z| = 2$. Note that $O_z \cap O_{z'} = \emptyset$ for all $z, z' \in Z$ with $z \neq z'$, since otherwise $v^+ z^+ \overrightarrow{C} z'^- xv \overleftarrow{C} z'x'z \overrightarrow{C} v^+$ is a cycle in $\mathscr{C}_{2l}$, where $x' \in O_z \cap O_{z'}$ and z appears in $v^+ \overrightarrow{C} z'$. Let $O^* := \bigcup_{z \in Z} O_z$ so $|O^*| = 2|Z|$. Since $\mathscr{C}_{2l} = \emptyset$, $(N_H(x) \cup N_H(v^+)) \cap O^* = \emptyset$ and $N_H(x) \cap N_H(v^+) = \emptyset$. Then $|C \cap B| = |C \cap A| \ge |N_C(v^+)^- \cup N_C(x)^+| = |N_C(v^+)| + |N_C(x)| - |Z| - 1$ and $|H \cap B| \ge |N_H(v^+)| + |N_H(x)| + |O^*|$. This means that

$$
\begin{aligned}
|B| &\ge \big(|N_C(v^+)| + |N_C(x)| - |Z| - 1\big) + \big(|N_H(v^+)| + |N_H(x)| + 2|O^*|\big) \\
&\ge d_G(v^+) + d_G(x) - 1 + |Z| \\
&\ge 2((|B| + 1)/2) - 1 + |Z| \ge |B|,
\end{aligned}
$$

so equality holds in the above inequalities. In particular, $Z = \emptyset$ and $C \cap A = N_C(v^+)^- \cup N_C(x)^+$. Let $M := N_H(v^+)$ so then $M \cap H^x = \emptyset$ and $H \cap B = M \cup H^x$.

If $M = \emptyset$, then $N_G(v^+) \subseteq V(C) \cap B$, and $2l - 2 = |C| \ge 2|N_G(v^+)| \ge 2\delta(A) \ge 2l + 2$, a contradiction. Hence, $M \neq \emptyset$.

Suppose that $N_C(x) - \{v\} \neq \emptyset$, say $v' \in N_C(x) - \{v\}$. By the symmetry of v and v', $N_H(v'^+) = H \cap B - H^x = M$. Let $w \in M$ so then $wv^+ \overrightarrow{C} v'xv \overleftarrow{C} v'^+w$ is a cycle in $\mathscr{C}_{2l}$, a contradiction. Hence, $N_C(x) = \{v\}$ and this implies that $C \cap B = N_C(v^+)$.

Therefore, $N_C(y) = \emptyset$ for any vertex $y \in H^x$, since otherwise, $v^+ \overleftarrow{C} uyxv \overrightarrow{C} u^+v^+$ is a cycle in $\mathscr{C}_{2l}$, where $u \in N_C(y)$, a contradiction. Since $v \in B$, v is not a cut vertex of G, and hence there exist vertices $y \in H^x$ and $x' \in N_H(y) - \{x\}$ such that $N_C(x') - \{v\} \neq \emptyset$ or $N_H(x') \cap M \neq \emptyset$. If $N_C(x') - \{v\} \neq \emptyset$, then, for

$v' \in N_C(x') - \{v\}$, let $C' := xyx'v'\overleftarrow{C}v^+v'^{+2}\overrightarrow{C}vx$. If $N_H(x') \cap M \neq \emptyset$, then, for $w \in N_H(x') \cap M$, let $C' := vxyx'wv^+v^{+4}\overrightarrow{C}v$. Then $C' \in \mathscr{C}_{2l}$, a contradiction. $\square$

This completes the proof of Lemma 3.10. $\square$

Lemma 3.11 ([9]) *Let $G = A \cup B$ be a disconnected bipartite graph. If $|A| \geq 2$, $|B| \geq 4$ and $\delta(A) \geq (|B| + 1)/2$, then $\mathscr{C}_{2l} \neq \emptyset$ for any $2 \leq l \leq \min\{|A|, \delta(A) - 1\}$.*

Proof Since $\delta(A) \geq (|B| + 1)/2$, there exists a component G' of G such that $A \subseteq V(G')$. Let $G' := A \cup B'$, where $B' \subseteq B$. Since $d_{G'}(a) \geq \delta(A) \geq (|B| + 1)/2 > (|B'| + 1)/2$ for any $a \in V(G') \cap A$, it follows from Lemma 3.9 that G' has no cut vertex in B'. Suppose that $|B'| \geq 4$. Then, by Lemma 3.10, $\mathscr{C}_{2l} \neq \emptyset$ for any $2 \leq l \leq \min\{|A|, \delta_{G'}(A) - 1\} = \min\{|A|, \delta(A) - 1\}$. Hence, we may assume that $|B'| \leq 3$. Then $5/2 \leq (|B| + 1)/2 \leq \delta(A) \leq |B'| \leq 3$, that is $\delta(A) = |B'| = 3$. This implies that G' is a complete bipartite graph, and $\min\{|A|, \delta(A) - 1\} = 2$ since $|A| \geq 2$. Therefore, $\mathscr{C}_{2l} \neq \emptyset$ for any $2 \leq l \leq \min\{|A|, \delta(A) - 1\}$. $\square$

We are finally able to prove Lemma 3.8.

Proof of Lemma 3.8. Let $G := A \cup B$ be a bipartite graph with $|A| \geq 2$, $|B| \geq 4$ and suppose $\delta(A) \geq (|B| + 1)/2$. Let l be an integer with $2 \leq l \leq \min\{|A|, \frac{|B|-1}{2}\}$. If G is disconnected, or if G is connected and has no cut vertex in B, then it follows from Lemmas 3.10 and 3.11 and the fact $l \leq \frac{|B|-1}{2} \leq \delta(A) - 1$ that G contains a copy of C_{2l}. If G is connected and has a cut vertex in B, then by Lemma 3.9, G is isomorphic to G_0. $\square$

In stead of proving Theorem 3.30, we prove the following lemma, which directly implies Theorem 3.30.

Lemma 3.12 ([9]) *Let n be an integer with $n \geq 2$. Let G be a Gallai colored complete graph. Let $k'(G)$ be the number of colors that induce a graph having a component of order at least n in G. If $|G| \geq (n - 1)k'(G) + 3n$, then G has a monochromatic C_{2n}.*

Proof Let $k' := k'(G)$. If $k' = 0$, this contradicts Lemma 3.2, so we may assume that $k' \geq 1$. We consider the following two cases, depending on k'.

Case 1 $k' = 1$.

Then $|G| \geq 4n - 1$. Since $k' = 1$, all but one color (say red) induces a graph which is not connected. Then it follows from Lemma 3.2 that there exists a Gallai partition $\mathscr{V}$ of G using only red edges between the parts. Choose such a partition $\mathscr{V}$ so that $|\mathscr{V}|$ is as large as possible.

We shall prove that each set of $\mathscr{V}$ has order at most $n - 1$. Let $V \in \mathscr{V}$ and suppose that $|V| \geq n$. It follows from Lemma 3.2 that V also has a Gallai partition such that for any color i on the edges between the parts, the subgraph of the reduced graph induced by i is connected. Since $k' = 1$ and $|V| \geq n$, the Gallai partition of V has only one color, which is red, on edges between the parts. But, adding the partition of V into $\mathscr{V}$ instead of V, this contradicts the maximality of $|\mathscr{V}|$, and hence $|V| \leq n - 1$. Therefore, we can see that each set of $\mathscr{V}$ has order at most $n - 1$.

Choose a bipartition of the vertices of G into A and B so that each of A and B is made up of sets of $\mathscr{V}$ and the orders of A and B are as balanced as possible. Since all the sets have order at most $n - 1$, we know that $||A| - |B|| \leq n - 1$. Since $|G| \geq 4n - 1$, this implies that $|A|, |B| \geq \frac{3n}{2} \geq n$. Thus, it is easy to find a red C_{2n} using only edges between A and B, completing the proof in this case.

Case 2 $k' \geq 2$.

We prove this case by induction on k'. Let $D_1, D_2, \ldots, D_{k'} \subseteq V(G)$ (say $D := \bigcup_{i=1}^{k'} D_i$) be disjoint sets of vertices such that each vertex $x \in D_i$ has only the color i on the edges from x to $G - D$, and $|G - D| \geq n$. We take such sets $D_1, D_2, \ldots, D_{k'}$ so that $|D|$ is as large as possible.

Claim 3.33 $|V| \leq n - 1$ for each $V \in \{D_1, D_2, \cdots, D_{k'}\}$.

Proof Suppose that $|D_i| \geq n$ for some $1 \leq i \leq k'$. Since $|G - D| \geq n$, the graph induced on $D_i \cup (G - D)$ contains a monochromatic $K_{n,n}$, which contains a monochromatic C_{2n} colored by i. Thus, $|D_i| \leq n - 1$ for each color i. $\square$

Now consider a Gallai partition of $G - D$ as in Lemma 3.2, and let $V_1, V_2, \ldots, V_t$ be the parts of $G - D$ under the Gallai partition. Then any color i on the edges between the parts in this partition induces a connected graph in $G - D$. Let

$$\mathscr{V} := \{D_1, D_2, \cdots, D_{k'}\} \cup \{V_1, V_2, \cdots, V_t\}.$$

Claim 3.34 Let e be an edge with color i between V and V' for some $V, V' \in \mathscr{V}$ with $V \neq V'$. Then $i \in \{1, 2, \ldots, k'\}$. In particular, if e is an edge between D_j and $D_{j'}$ for some $1 \leq j < j' \leq k'$, then $i = j$ or $i = j'$.

Proof Let e be an edge with color i between V and V' for some $V, V' \in \mathscr{V}$ with $V \neq V'$.

Suppose first that $V, V' \in \{V_1, V_2, \ldots, V_t\}$ and $i \notin \{1, 2, \ldots, k'\}$. Since any color on the edges between the parts in the partition of $G - D$ induces a connected graph, i is also connected in the reduced graph of the partition. Since $i \notin \{1, 2, \ldots, k'\}$, $G - D$ has at most $n - 1$ vertices. Then, by Claim 3.33, $(n - 1)k' + 3n \leq |V(G)| \leq (n - 1)k' + (n - 1)$, a contradiction. So, if $V, V' \in \{V_1, V_2, \ldots, V_t\}$, then $i \in \{1, 2, \ldots, k'\}$.

Suppose next that $V \in \{D_1, D_2, \cdots, D_{k'}\}$ and $V' \in \{V_1, V_2, \cdots, V_t\}$. In this case, all edges between V and V' is colored by j, where $V = D_j$. Then $i = j \in \{1, 2, \cdots, k'\}$.

Suppose finally that $V = D_j$ and $V' = D_{j'}$ for some $1 \leq j < j' \leq k'$ and $i \neq j, j'$. Let $e := x_j x_{j'}$ with $x_j \in D_j$ and $x_{j'} \in D_{j'}$. Taking any vertex $y \in V(G) - D$, the edges $e, x_i y$ and $x_j y$ are colored by i, j and j', respectively, the triangle $x_i x_j y$ is rainbow, a contradiction. $\square$

Claim 3.35 $|V| \leq n - 1$ for each $V \in \{V_1, V_2, \cdots, V_t\}$.

Proof If $|V_i| \geq n$ for some $1 \leq i \leq t$, then, by Claim 3.34, we can add $(G - D) - V_i$ into D, contradicting the maximality of $|D|$.   □

Subcase 2.1 $k' = 2$.

In this case, note that $|V(G)| \geq 5n - 2$. We take a subset $B' \subseteq V(G)$ so that B' consists of the union of sets in $\mathcal{V}$ with the additional restriction that $|B'| \geq 2n + 1$. Choose such a set B' so that $|B'|$ is as small as possible. By Claims 3.33 and 3.35, $|V| \leq n - 1$ for each $V \in \mathcal{V}$, and hence the choice of a smallest set B' implies that $|B'| \leq 3n - 1$. Hence, we have

$$|V(G) - B'| \geq 5n - 2 - (3n - 1) = 2n - 1.$$

When $|B'|$ is odd, then let $B := B'$; otherwise let $B := B' - \{u\}$ for some u. We construct B so that $|B|$ is odd in either case.

For $i = 1, 2$, let

$$A^i := \{x \in V(G) - B' : x \text{ has at least } \frac{|B| + 1}{2} \text{ edges to } B \text{ with color } i.\}.$$

By Claim 3.34, all edges between B' and $V(G) - B'$ are colored by either color 1 or color 2. Hence, we have that $V(G) - B' = A^1 \cup A^2$. We may assume that $|A^1| \geq \left\lceil \frac{|V(G) - B'|}{2} \right\rceil \geq n$.

Let H be the bipartite subgraph of G on $A^1 \cup B$ induced by the edges colored by color 1. Then $H = A^1 \cup B$ is a bipartite graph with $\delta_B(A^1) \geq \frac{|B|+1}{2}$, and $|B| \geq 2n + 1 \geq 5$. Hence, by Lemma 3.8, either H contains a C_{2l} for each $2 \leq l \leq \min \left\{ |A^1|, \frac{|B|-1}{2} \right\}$ or H is isomorphic to G_0. If the former holds, then in particular, H contains a C_{2n} since $\min \left\{ |A^1|, \frac{|B|-1}{2} \right\} \geq n \geq 2$. Thus, $C_{2n} \subseteq H \subseteq G$ so G contains the desired cycle colored by 1. Therefore, we may assume that H is isomorphic to G_0.

Let $H := B_1 + A_1 + B_2 + A_2 + B_3$ with $A_1, A_2 \neq \emptyset$, $|B_1| = |B_3|$ and $|B_2| = 1$. Note that $|B_1| = |B_3| = \frac{|B|-1}{2} \geq n$.

Let $\widetilde{A_1} := \{x \in A^1 \cup A^2 : \text{all edges between } x \text{ and } B_3 \text{ have color } 2\}$,

and $\widetilde{A_2} := V(G) - B' - \widetilde{A_1}$.

Since H has no edges between A_1 and B_3, we have that $A_1 \subseteq \widetilde{A_1}$. Note that for all $x \in A_2$, we have that $x \in \widetilde{A_2}$ and all edges between x and B_1 are colored by 2. If there exist two edges xy_1 and xy_3 colored by 1 such that $x \in A^2$, $y_1 \in B_1$ and $y_3 \in B_3$, then, together with x and the two edges xy_1 and xy_3, we can find a monochromatic C_{2n} colored by 1 in H. Thus, we may assume that for all $x \in A^2$, all edges between x and B_j are colored by 2 for some $j = 1, 3$. Hence, we also have that for each $x \in \widetilde{A_2}$, all edges between x and B_1 have color 2.

Suppose first that $|\widetilde{A_1}| \geq n$. Then with $\widetilde{A_1}$ and B_3 forming a complete bipartite graph with all edges colored in color 2 and $|\widetilde{A_1}|, |B_3| \geq n$, we can find a monochromatic C_{2n} colored by 2. Thus, we may assume that $|\widetilde{A_1}| < n$. Since $\widetilde{A_1} \cup \widetilde{A_2} = V(G) - B'$ has at least $2n - 1$ vertices, we have that $|\widetilde{A_2}| \geq n$. In this case, with $\widetilde{A_2}$ and B_1 forming a complete bipartite graph with all edges having color 2 and $|\widetilde{A_2}|, |B_1| \geq n$, we are also done, completing the proof in this case.

Subcase 2.2 $k' \geq 3$.

Since there are at most two colors between the parts of $G - D$, there is an integer i with $1 \leq i \leq k'$, by symmetry say $i = k'$, such that the color k' does not appear on the edges between different parts in $G - D$. Let $G' := G - D_{k'}$. By Claim 3.33,

$$|G'| \geq |G| - |D_{k'}|$$
$$\geq (n - 1)(k' - 1) + 3n.$$

By Claim 3.34, for every $1 \leq p < q \leq k' - 1$, every edge between D_p and D_q is not colored by k'. This and the choice of $i = k'$ implies that every component of the subgraph induced on color k' in G' is contained in some $V \in \mathcal{V}$. Since $|V| \leq n - 1$, the color k' has no component in G' of order at least n. This implies that $k'(G') \leq k'(G) - 1$, and by the induction hypothesis, G' has a monochromatic C_{2n}. Since $G' \subseteq G$, G has a monochromatic C_{2n}. This completes the proof of Lemma 3.12 and thereby the proof of Theorem 3.30. $\square$

In keeping with bipartite graphs, we next consider bounds for paths by Hall et al.

Theorem 3.36 ([9]) *For all integers k and n with $k \geq 1$ and $n \geq 3$,*

$$\left\lfloor \frac{n-2}{2} \right\rfloor k + \left\lceil \frac{n}{2} \right\rceil + 1 \leq gr_k(K_3 : P_n) \leq \left\lfloor \frac{n-2}{2} \right\rfloor k + 3 \left\lfloor \frac{n}{2} \right\rfloor.$$

In light of Theorem 3.30, the proof of Theorem 3.36 becomes relatively easy. Note that the improvement of Theorem 3.30 by Zhang et al. [20] implies an improvement of Theorem 3.36 as well.

Proof Let G be a Gallai colored complete graph of order at least $\left\lfloor \frac{n-2}{2} \right\rfloor k + 3 \left\lfloor \frac{n}{2} \right\rfloor$ and suppose, for a contradiction, that G contains no monochromatic P_n. By Theorem 3.30, there exists a monochromatic copy of C_m with $m = 2 \left\lfloor \frac{n}{2} \right\rfloor$. If n is even, this cycle contains a monochromatic copy of P_n. Hence, we may assume that n is odd.

For convenience, let t be chosen so that $2t + 1 = n$ and note that we are assuming that G contains no monochromatic copy of P_{2t+1}. Let k be the color of the edges of C_m. This choice implies that any edge between a vertex in $V(G) \setminus V(C)$ and a vertex in C cannot have the color k. If there exists a vertex $x \in V(G) \setminus V(C)$ such that x has at least two colors on the edges from x to $V(C)$, then two of the colors together with k induces a rainbow triangle, a contradiction. Thus, each vertex $x \in V(G) \setminus V(C)$

has only one color on the edges from x to $V(C)$. Therefore, $V(G) \setminus V(C)$ can be partitioned into $D_1, D_2, \ldots, D_{k-1}$ so that each vertex $x \in D_i$ has only one color i on the edges from x to $V(C)$. Since there exists no monochromatic P_{2t+1} and $|C| = 2t$, we see that $|D_i| \leq t - 1$ for any $1 \leq i \leq k - 1$. Hence,

$$(t - 1)k + 3t \leq |V(G)| \leq |V(C)| + \sum_{1 \leq i \leq k-1} |D_i| \leq 2t + (t - 1)(k - 1),$$

a contradiction.   □

3.4.2 Complete Bipartite Graphs

For the complete bipartite graphs $K_{3,m}$, the following was proven by Wu et al. [1].

Theorem 3.37 ([1]) *For $k \geq 3$ and $m \geq 3$, we get*

$$R(K_{3,m}, K_{3,m}) + 2(k - 2) \leq gr_k(K_3 : K_{3,m})$$
$$\leq \max\{(6m - 2), R(K_{3,m}, K_{3,m})\} + 2(k - 2).$$

In particular, for the complete bipartite graph $K_{3,3}$, this reduces down to the exact Gallai-Ramsey number.

Corollary 3.2 ([1]) *For $k \geq 3$, we have*

$$gr_k(K_3 : K_{3,3}) = 2k + 14.$$

Since $2^{(3m-1)/(3+m)} \leq R(K_{3,m}, K_{3,m}) \leq 8m - 2$ from [34] and [35], respectively, the sharpness of this result in general depends on the yet unknown 2-color Ramsey number for $K_{3,m}$. The only small value of $m \geq 3$ for which $R(K_{3,m}, K_{3,m})$ is known is when $m = 3$ (see [36]). Since $R(K_{3,3}, K_{3,3}) = 18 \geq 6(3) - 2$, the bounds in Theorem 3.37 are equal, establishing the conclusion of Corollary 3.2 as noted above. Otherwise, the general relationship between $R(K_{3,m}, K_{3,m})$ and $6m - 2$ remains unclear.

We include a modified version of the proof by Wu et al. [1].

Proof *(Proof of Theorem 3.37)* The lower bound follows from Proposition 3.1. For the upper bound, suppose G is a Gallai coloring of K_n using at most k colors where

$$n = \max\{(6m - 2), R(K_{3,m}, K_{3,m})\} + 2(k - 2)$$

and suppose G contains no monochromatic copy of $K_{3,m}$. By Theorem 3.2, we may consider a 3-coloring G' of $K_{n-2(k-2)}$ with a Gallai partition with colors red and blue in which all parts have order at most 2. In particular, since $|G'| \geq 6m - 2$, this means that there are at least $3m - 1$ parts in this Gallai partition.

Since $|G'| \geq R(K_{3,m}, K_{3,m})$, there must exist at least one part A of this Gallai partition of order 2. First suppose there is a color, say red, with at most $2m - 1$ vertices in $G' \setminus A$ having red edges to A. This means that there are at least

$$n - 2(k - 2) - 2 - (2m - 1) \geq 4m - 3$$

vertices in $G' \setminus A$ with all blue edges to A. Let B be a set of $4m - 3$ of these vertices. Let v_1, v_2, v_3 be three of the vertices in $G' \setminus (A \cup B)$. In order to avoid creating a blue copy of $K_{3,m}$, the vertex v_i can have at most $m - 1$ blue edges to B so all remaining edges must be red. With $|B| = 4m - 3$, there must be a set of at least m vertices in B with all red edges to v_1, v_2, v_3, creating a red copy of $K_{3,m}$. This means that there is no color c (among red and blue) for which at most $2m - 1$ vertices in $G' \setminus A$ have color c on edges to A.

Let B (and C) be the set of vertices in $G \setminus A$ with blue (respectively, red) edges to A. Note that we must have $|B|, |C| \geq 2m$. In order to avoid creating a red $K_{3,m}$, each vertex in B must have at most $m - 1$ red edges to C. Symmetrically, each vertex in C must also have at most $m - 1$ blue edges to B. This means that there are a total of at most $(|B| + |C|)(m - 1)$ edges between B and C but since the graph is complete, we know there are exactly $|B| \cdot |C|$ edges between B and C. Since $|B|, |C| \geq 2m$, we get

$$(|B| + |C|)(m - 1) < |B|m + |C|m \leq \frac{|B||C|}{2} + \frac{|B||C|}{2} = |B||C|,$$

a contradiction, completing the proof.   □

For larger complete bipartite graphs, the following bounds have been shown by Chen et al.

Theorem 3.38 ([37]) *For fixed integers $k \geq 2$ and $m \geq 1$, if $\ell \to \infty$, then*

$$(1 - o(1))2^m n \leq gr_k(K_3 : K_{\ell,m}) \leq (2^m + 2^{m/2+1} + k)n + 4m^3.$$

Using this result, we can obtain the following bounds on the Gallai-Ramsey numbers for all complete bipartite graphs.

Theorem 3.39 ([1]) *Given positive integers ℓ, m where $\ell \leq m$, let $H = K_{\ell,m}$ and let $R = \max\{R(H, H), 3b(H) - 2\}$. Then*

$$R + (\ell - 1)(k - 2) \leq gr_k(K_3 : H) \leq (R + k - 3)(\ell - 1) + 1.$$

We include the easy proof by Wu et al. [1].

Proof The lower bound follows from Proposition 3.1. For the upper bound, suppose G is a Gallai coloring of K_n using at most k colors where $n = (R - 1)(\ell - 1) + (\ell - 1)(k - 2)$ and suppose G contains no monochromatic copy of $K_{\ell,m}$.

By Theorem 3.2, we may consider a 3-coloring G' of $K_{(R-1)(\ell-1)+1}$ with a Gallai partition in which all parts have order at most $\ell - 1$. By the definition of R, there are at most $R - 1$ parts so with each part having order at most $\ell - 1$, there can be at most $(R - 1)(\ell - 1)$ vertices in G', a contradiction.   □

3.4.3 Books

The *book* graph with m *pages* is denoted by B_m, where $B_m = K_2 + \overline{K_m}$. See Fig. 3.13 for a drawing of B_4. Call the central edge (the K_2) the *spine* of the book. Note that $B_1 = K_3$ and $B_2 = K_4 \setminus \{e\}$ where e is an edge of the K_4. In this work, we prove bounds on the Gallai-Ramsey number of all books, with sharp results for several small books. Since books are not bipartite, in light of Theorem 3.1, it should come as no surprise that all of our results are exponential as a function of the number of colors k.

Theorem 3.40 *Given integers $k \geq 1$ and $2 \leq m \leq 5$,*

$$gr_k(K_3 : B_m) = \begin{cases} m + 2 & \text{if } k = 1, \\ (R_m - 1) \cdot 5^{(k-2)/2} + 1 & \text{if } k \text{ is even}, \\ 2 \cdot (R_m - 1) \cdot 5^{(k-3)/2} + 1 & \text{if } k \geq 3 \text{ is odd}. \end{cases}$$

Call a color *m-admissible* if it induces a subgraph with maximum degree at least m, and *m-inadmissible* otherwise. We begin this section with a lemma by Zou et al. about the case when only one color is admissible.

Lemma 3.13 ([38]) *Given integers $m \geq 2$ and $k \geq 2$, let n be the largest number of vertices in a k-coloring of a complete graph in which there is*

- *no rainbow triangle,*
- *no monochromatic copy of B_m,*
- *a Gallai partition with all parts having order at most $m - 1$ and*
- *only one m-admissible color.*

Then

$$n \leq \begin{cases} 3m - 1 & \text{if } k = 2, \\ 5m - 5 & \text{otherwise}. \end{cases}$$

Proof First suppose $k = 2$ and let G be a 2-coloring of K_{3m} with no monochromatic copy of B_m, supposing that only one of the two colors is m-admissible. Note that G trivially contains no rainbow triangle since it uses only 2 colors. Say red is the

Fig. 3.13 The book graph B_4

m-admissible color and blue is the other (m-inadmissible) color. Let u and v be any two vertices of G that are joined by a red edge. Then there are $3m - 2$ other vertices in G but at most $m - 1$ of them can have blue edges to u and at most $m - 1$ of them can have blue edges to v. This leaves at least

$$(3m - 2) - 2(m - 1) \geq m$$

vertices with red edges to both u and v, creating a red copy of B_m, for a contradiction.

Now suppose $k \geq 3$ and let G be a k-coloring of K_{5m-4} with no monochromatic B_m and suppose only one of the colors is m-admissible. Consider a Gallai partition with all parts having order at most $m - 1$ as provided by the hypothesis. Say red is the m-admissible color, so red must appear in the Gallai partition, and let blue be the other color appearing in the Gallai partition. Choose two parts of this partition with red edges between them, say H_1 and H_2. Then since $|G| = 5m - 4$, there are at least $3m - 2$ remaining vertices in $G \setminus (H_1 \cup H_2)$. At most $m - 1$ of these remaining vertices can have blue edges to H_1 and at most $m - 1$ of these vertices can have blue edges to H_2. This means that at least m vertices must have red edges to both H_1 and H_2, producing a red copy of B_m, for a contradiction. $\qquad \square$

Let $R_m = R(B_m, B_m)$ and define

$$R'_m = \sum_{i=1}^{m-1} [R_{\lceil m/i \rceil} - 1].$$

This quantity provides a bound on a type of restricted Ramsey number as seen in the following lemma.

Lemma 3.14 *For $m \geq 2$, the largest number of vertices in a Gallai coloring of a complete graph containing no monochromatic copy of B_m in which all parts of the Gallai partition have order at most $m - 1$ is at most R'_m.*

Proof Suppose, for a contradiction, that G is a k-coloring of K_n containing no rainbow triangle and no monochromatic copy of B_m where

$$n = \sum_{i=1}^{m-1} [R_{\lceil m/i \rceil} - 1] + 1,$$

and suppose all parts of a Gallai partition of G have order at most $m - 1$. There are certainly at most $R_m - 1$ parts in this Gallai partition since otherwise the reduced graph would contain a monochromatic copy of B_m. Similarly, there are at most $R_{\lceil m/2 \rceil} - 1$ parts of order at least 2 in the Gallai partition since otherwise the reduced graph of this subset of parts would contain a monochromatic copy of $B_{\lceil m/2 \rceil}$, which implies the existence of a monochromatic copy of B_m in G.

More generally, for each integer i with $1 \leq i \leq m - 1$, there are at most $R_{\lceil m/i \rceil} - 1$ parts of order at least i in the Gallai partition since otherwise the reduced graph of this subset of parts would contain a monochromatic copy of $B_{\lceil m/i \rceil}$, which creates a monochromatic copy of B_m in G. Summing over the values of i, we see that

$$|G| = n \leq \sum_{i=1}^{m-1} [R_{\lceil m/i \rceil} - 1] < n,$$

a contradiction, completing the proof of Lemma 3.14.   □

With one more definition, we may state the general upper bound for books. Let $\ell = \ell(m)$ be the number of colors that are m-inadmissible and define the quantity $gr_{k,\ell}(K_3 : H)$ to be the minimum integer n such that every k coloring of K_n, with at least ℓ different m-inadmissible colors, contains either a rainbow triangle or a monochromatic copy of H.

Theorem 3.41 ([38]) *Given positive integers $k \geq 1$, $m \geq 3$ and $0 \leq \ell \leq k$, let*

$$gr_{k,\ell,m} = \begin{cases} m + 2 - \ell & \text{if } k = 1, \\ R'_m \cdot 5^{\frac{k-2}{2}} + 1 - (m-1)\ell & \text{if } k \text{ is even}, \\ 2 \cdot R'_m \cdot 5^{\frac{k-3}{2}} + 1 - (m-1)\ell & \text{if } k \geq 3 \text{ is odd}. \end{cases}$$

Then

$$gr_{k,\ell}(K_3 : B_m) \leq gr_{k,\ell,m}.$$

Note that this implies that

$$\begin{cases} (4 + o(1))m \cdot 5^{(k-2)/2} + 1 & \text{if } k \text{ is even}, \\ 2 \cdot (4 + o(1))m \cdot 5^{(k-3)/2} + 1 & \text{if } k \text{ is odd} \end{cases}$$

$$\leq gr_k(K_3 : B_m)$$

$$\leq \begin{cases} m + 2 - \ell & \text{if } k = 1, \\ (4 + o(1))m \ln[(4 + o(1))m] \cdot 5^{\frac{k-2}{2}} + 1 - (m-1)\ell & \text{if } k \text{ is even}, \\ 2 \cdot (4 + o(1))m \ln[(4 + o(1))m] \cdot 5^{\frac{k-3}{2}} + 1 - (m-1)\ell & \text{if } k \geq 3 \text{ is odd}. \end{cases}$$

Proof For a contradiction, suppose G is a k-coloring of K_n with

$$n = gr_{k,\ell,m}$$

containing no rainbow triangle and no monochromatic copy of B_m with the property that there are at least ℓ different m-inadmissible colors in G. We prove the result by induction on $k - \ell$. If $k = 1$, then Theorem 3.41 is immediate so next suppose $k = 2$. Then for any $m \geq 3$, it is easy to verify that

$$n = R'_m + 1 - \ell(m-1) \geq R'_m + 1 - 2(m-1) \geq R_m,$$

meaning that G contains a monochromatic B_m, for a contradiction. In particular, if $m = 3$, we have $n = 19 > 14 = R_3$ and if $m = 4$, we have $n = 29 > 18 = R_4$.

For the remainder of the proof, we may assume $k \geq 3$, and so $n \geq 2R'_m - 3(m - 1)$. If $k - \ell = 0$, then there are no m-admissible colors but with

$$n \geq 2R'_m - 3(m - 1) \geq 5m/2,$$

by Theorem 3.7, there is a vertex with at least $\frac{2n}{5} > m$ edges in a single color, so this color would be m-admissible, a contradiction.

Next suppose $k - \ell = 1$, and consider a Gallai partition. If there are only two parts, then either both parts have at least m vertices or one part has fewer than m vertices. If both parts have at least m vertices, then neither part contains any edges in this color, reducing k by 1 and reducing the problem to the case $k - \ell = 0$. Applying induction on $k - \ell$ within a largest part yields the desired result. On the other hand, if one part has fewer than m vertices, the vertices in this smaller part may be removed, yielding a graph with one more m-inadmissible color within the larger part, again reducing the problem to the case $k - \ell = 0$. More generally, if there are more than two parts, none of the parts can have order at least m since the only color with edges to a part of order at least m must be m-admissible, yielding a bipartition as above. With all parts having order at most $m - 1$, but only one m-admissible color, Lemma 3.13 implies the existence of a monochromatic B_m in the m-admissible color, to complete the case $k - \ell = 1$.

This means we may assume $k - \ell \geq 2$. There is a Gallai partition of G, say using colors red and blue. Consider such a Gallai partition with the smallest number of parts, say t. If $t \geq R_m$, then the reduced graph of this partition contains a monochromatic copy of B_m, a contradiction. We may therefore assume $t \leq R_m - 1$. Let $H_1, H_2, \ldots, H_t$ be parts of such a Gallai partition.

First suppose $t \leq 3$. If $t = 3$, then the reduced graph is a triangle but this contains a bipartition with only one color on the edges between the parts so we may assume $t = 2$. Let red be the color of the edges between the two parts. If $|H_1| \leq m - 1$, then red is m-inadmissible within H_2 so we remove H_1 from the graph and apply induction on $k - \ell$ within H_2. This means that $|H_1|, |H_2| \geq m$, but then there can be no red edges within either H_1 or H_2. By induction on $k - \ell$, this gives

$$
\begin{aligned}
n &= |H_1| + |H_2| \\
&\leq 2\left[gr_{k-1,\ell,m} - 1\right] \\
&< gr_{k,\ell,m},
\end{aligned}
$$

a contradiction. We may therefore assume $t \geq 4$ and, by Lemma 3.2, each part of the Gallai partition has incident edges in both colors that appear in the Gallai partition. Suppose red and blue are the two colors appearing in the Gallai partition.

If one of red or blue was m-inadmissible in G, then since both colors must induce a connected subgraph of the reduced graph (by Lemma 3.2), all parts must have order less than m, so by Lemma 3.14, we have $|G| \leq R'_m$. This is a contradiction since $k \geq 3$, so red and blue must both be m-admissible within G.

Let r be the number of parts of the Gallai partition with order at least m, say with

$$|H_1|, |H_2|, \ldots, |H_r| \geq m \quad \text{and} \quad |H_{r+1}|, |H_{r+2}|, \ldots, |H_t| \leq m - 1.$$

If $r \geq 4$ and $t \geq 6$, then any choice of 6 parts containing the 4 parts $\mathscr{H} = \{H_1, H_2, H_3, H_4\}$ will contain a monochromatic triangle in the reduced graph. Such a triangle must contain at least one part from $\mathscr{H}$, meaning that the corresponding subgraph of G must contain a monochromatic copy of B_m, for a contradiction. Thus, we may assume either $4 \leq t \leq 5$ or $r \leq 3$. We consider cases based on the value of r. Note that the case $r = 0$ follows from Lemma 3.14.

Case 1 $r = 1$.

Since $t \geq 4$, both red and blue are m-inadmissible within H_1 and if $|H_1| \geq n - 2(m - 1)$, we may simply apply induction on $k - \ell$, so suppose $|H_1| \leq n - 2(m - 1) - 1$. This means that there are at least m vertices in $G \setminus H_1$ with the same color on all edges to H_1, say red. This implies that there are no red edges within H_1 so $|H_1| \leq gr_{k-1,\ell+1,m} - 1 < \frac{n}{2}$.

Let H_r (and H_b) be the subsets of vertices of $G \setminus H_1$ that have all red (or blue, respectively) edges to H_1. To avoid a red B_m, all edges between the parts contained in H_r must be blue and similarly all edges between the parts contained in H_b must be red. At this point, we observe an easy fact.

Fact 3.42 *For any $m \geq 3$ and $n' \geq 3m - 2$, in any Gallai coloring of a $K_{n'}$ in which only one color appears between the parts and all parts have order at most $m - 1$, there is a monochromatic copy of B_m.*

Therefore, by Fact 3.42, we have $|H_r|, |H_b| \leq 3m - 3$, so $|G \setminus H_1| \leq 6m - 6$ and

$$gr_{k,\ell,m} - (gr_{k-1,\ell+1,m} - 1) \leq 6m - 6.$$

More specifically, if $|H_b| \geq m$, then $|H_1| < gr_{k-2,\ell,m}$ so

$$gr_{k,\ell,m} - (gr_{k-2,\ell,m} - 1) \leq 6m - 6,$$

which means that $k \leq 2$, a case that has already been considered. On the other hand, if $|H_b| \leq m - 1$, then

$$gr_{k,\ell,m} - (gr_{k-1,\ell+1,m} - 1) \leq 4m - 4.$$

This again implies that $k \leq 2$, a case that has already been considered, completing the proof in this case.

Case 2 $r = 2$.

Suppose red is the color of the edges between H_1 and H_2 so neither H_1 nor H_2 can contain red edges and recall that blue must be m-inadmissible within H_1 and H_2. If there are at most $2(m - 1)$ vertices in $G \setminus (H_1 \cup H_2)$, then it is easy to show that $n < gr_{k,\ell}(K_3 : B_m)$ for a contradiction, so we may assume $|G \setminus (H_1 \cup H_2)| \geq 2(m - 1) + 1$.

For $i \in \{1, 2\}$, let A_i be the set of vertices in $G \setminus (H_1 \cup H_2)$ with red edges to H_i. Note that $A_1 \cap A_2 = \emptyset$ since any intersection would create a red copy of B_m. Then for $i \in \{1, 2\}$, let $C_i = G \setminus (H_1 \cup H_2 \cup A_{3-i})$, so C_i has all edges to H_{3-i} in blue. Since $|G \setminus (H_1 \cup H_2)| \geq 2(m - 1) + 1$, at least one of C_1 or C_2 contains at least m vertices, say $|C_1| \geq m$. Then H_2 contains no blue edges.

To avoid a blue copy of B_m, for each $i \in \{1, 2\}$, there can be no blue edges within C_i, meaning that all edges between the parts of the Gallai partition within C_i are red, so by Fact 3.42, we know $|C_i| \leq 4m - 6$ for each $i \in \{1, 2\}$. On the other hand, if $|C_2| \geq m$, then H_1 also contains no blue edges and must therefore be smaller. This means that

$$
\begin{aligned}
|G| &\leq |H_1| + |H_2| + |C_1 \cup C_2| \\
&\leq [gr_{k-1,\ell+1,m} - 1] + [gr_{k-2,\ell,m} - 1] + (4m - 6) + (m - 1) \\
&< gr_{k,\ell,m},
\end{aligned}
$$

a contradiction for $k \geq 3$, completing the proof in this case.

Case 3 $r = 3$.

To avoid a monochromatic B_m, the triangle in the reduced graph corresponding to the parts $\{H_1, H_2, H_3\}$ must not be monochromatic. Without loss of generality, suppose the edges from H_1 to H_2 are red and all other edges between these parts are blue. Then H_1 and H_2 contain no red or blue edges while H_3 contains no blue edges and red is m-inadmissible within H_3.

If there is a vertex $v \in G \setminus (H_1 \cup H_2 \cup H_3)$ with blue edges to H_3, then to avoid creating a blue B_m, v must have all red edges to both H_1 and H_2, producing a red B_m. This means that all edges between H_3 and $G \setminus (H_1 \cup H_2 \cup H_3)$ must be red. If $|G \setminus (H_1 \cup H_2 \cup H_3)| \leq m - 1$, then

$$
\begin{aligned}
|G| &\leq |H_1| + |H_2| + |H_3| + (m - 1) \\
&\leq 2[gr_{k-2,\ell,m} - 1] + [gr_{k-1,\ell+1,m} - 1] + (m - 1) \\
&< gr_{k,\ell,m},
\end{aligned}
$$

a contradiction, so suppose $|G \setminus (H_1 \cup H_2 \cup H_3)| \geq m$. Then H_3 contains no red edges and all edges between parts within $G \setminus (H_1 \cup H_2 \cup H_3)$ must be blue. By Fact 3.42, we get

$$|G| \leq |H_1| + |H_2| + |H_3| + (4m - 6)$$
$$\leq 3[gr_{k-2,\ell,m} - 1] + (4m - 6)$$
$$< gr_{k,\ell,m},$$

again a contradiction, completing the proof in this case.

Case 4 $r \geq 4$.

As observed previously, this implies that $4 \leq t \leq 5$. Looking only at the subgraph of the reduced graph induced on the r large parts, there can be no monochromatic triangle. If $r = 5$, there is only one coloring of K_5 with no monochromatic triangle and if $r = 4$, there are two colorings of K_4 with no monochromatic triangle. In all of these colorings, every vertex has an incident edge in both colors, meaning that all of the r corresponding parts of order at least m must have no red or blue edges. Then

$$|G| \leq \sum_{i=1}^{t} |H_i|$$
$$\leq 5[gr_{k-2,\ell,m} - 1]$$
$$< gr_{k,\ell,m},$$

a contradiction, completing the proof of this case, and therefore the proof of Theorem 3.41.   □

References

1. H. Wu, C. Magnant, P. Salehi Nowbandegani, S. Xia, All partitions have small parts—Gallai-Ramsey numbers of bipartite graphs. Discrete Appl. Math. **254**, 196–203 (2019)
2. C. Magnant, A general lower bound on Gallai-Ramsey numbers for non-bipartite graphs. Theo. Appl. Graphs **5**(1), Article 4 (2018)
3. A. Gyárfás, G. Sárközy, A. Sebő, S. Selkow, Ramsey-type results for gallai colorings. J. Graph Theory **64**(3), 233–243 (2010)
4. M. Axenovich, P. Iverson, Edge-colorings avoiding rainbow and monochromatic subgraphs. Discrete Math. **308**(20), 4710–4723 (2008)
5. F.R.K. Chung, R.L. Graham, Edge-colored complete graphs with precisely colored subgraphs. Combinatorica **3**(3–4), 315–324 (1983)
6. R.J. Faudree, R. Gould, M. Jacobson, C. Magnant, Ramsey numbers in rainbow triangle free colorings. Australas. J. Combin. **46**, 269–284 (2010)
7. A. Gyárfás, G. Simonyi, Edge colorings of complete graphs without tricolored triangles. J. Graph Theory **46**(3), 211–216 (2004)
8. H. Wu, C. Magnant, Gallai-Ramsey numbers for monochromatic triangles or 4-cycles. Graphs Combin. **34**(6), 1315–1324 (2018)
9. M. Hall, C. Magnant, K. Ozeki, M. Tsugaki, Improved upper bounds for Gallai-Ramsey numbers of paths and cycles. J. Graph Theory **75**(1), 59–74 (2014)
10. S.A. Burr, *Either a Graph or its Complement Contains a Spanning Broom*. Unpublished manuscript

11. A. Bialostocki, P. Dierker, W. Voxman, *Either a Graph or its Complement is Connected: A Continuing Saga*. Unpublished manuscript
12. C. Magnant, Z. Magnant, K. Ozeki, *On Gallai-Ramsey Numbers for Paths*. Manuscript
13. S. Fujita, C. Magnant, Gallai-Ramsey numbers for cycles. Discrete Math. **311**(13), 1247–1254 (2011)
14. J. Gregory, C. Magnant, Z. Magnant, *Gallai-Ramsey Number for the 8-Cycle*. Submitted
15. S. Fujita, C. Magnant, K. Ozeki, Rainbow generalizations of Ramsey theory—a dynamic survey. Theo. Appl. Graphs 0(1) (2014)
16. C. Bosse, Z. Song, *Multicolor Gallai-Ramsey numbers of C_9 and C_{11}*. Submitted
17. C. Bosse, Z. Song, J. Zhang, *Multicolor Gallai-Ramsey Numbers of c_{13} and c_{15}*. Submitted
18. D. Bruce, Z. Song, Gallai-Ramsey numbers of C_7 with multiple colors. Discrete Math. **342**(4), 1191–1194 (2019)
19. Z. Wang, Y. Mao, C. Magnant, I. Schiermeyer, J. Zou, *Gallai-Ramsey Numbers of Odd Cycles*. Submitted
20. F. Zhang, Y. Chen, Z. Song, *Gallai-Ramsey Numbers of Cycles*. Submitted
21. G.A. Dirac, Some theorems on abstract graphs. Proc. Lond. Math. Soc. 3(2), 69–81 (1952)
22. P. Erdős, T. Gallai, On maximal paths and circuits of graphs. Acta Math. Acad. Sci. Hungar **10**, 337–356 (1959)
23. R.J. Faudree, R.H. Schelp, All Ramsey numbers for cycles in graphs. Discrete Math. **8**, 313–329 (1974)
24. G. Károlyi, V. Rosta, Generalized and geometric Ramsey numbers for cycles. Theoret. Comput. Sci. **263**(1–2), 87–98 (2001). Combinatorics and computer science (Palaiseau, 1997)
25. V. Rosta, On a Ramsey-type problem of J.A. Bondy, and P. Erdős. I, II. J. Combinatorial Theory Ser. B **15**, 94–104; ibid. **15**, 105–120 (1973)
26. J. Fox, A. Grinshpun, J. Pach, The Erdős-Hajnal conjecture for rainbow triangles. J. Combin. Theory Ser. B **111**, 75–125 (2015)
27. H. Liu, C. Magnant, A. Saito, I. Schiermeyer, Y. Shi, *Gallai-Ramsey Number for K_4* (Accepted - J, Graph Theory, 2019)
28. C. Magnant, I. Schiermeyer, *Gallai-Ramsey Number for K_5*. Submitted
29. B.D. McKay, S.P. Radziszowski, Subgraph counting identities and Ramsey numbers. J. Combin. Theory Ser. B **69**(2), 193–209 (1997)
30. L.E. Bush, The William Lowell Putnam mathematical competition. Amer. Math. Monthly **60**(8), 539–542 (1953)
31. V. Chvátal, F. Harary, Generalized Ramsey theory for graphs. II. Small diagonal numbers. Proc. Am. Math. Soc. **32**, 389–394 (1972)
32. R.E. Greenwood, A.M. Gleason, Combinatorial relations and chromatic graphs. Canad. J. Math. **7**, 1–7 (1955)
33. Z. Wang, C. Magnant, Y. Mao, J. Zou, *Ramsey and Gallai-Ramsey Numbers for Two Classes of Unicyclic Graphs*. Submitted
34. T.K. Mishra, S.P. Pal, Lower bounds for Ramsey numbers for complete bipartite and 3-uniform tripartite subgraphs. J. Graph Algorithms Appl. **17**(6), 671–688 (2013)
35. R. Lortz, I. Mengersen, Bounds on Ramsey numbers of certain complete bipartite graphs. Results Math. **41**(1–2), 140–149 (2002)
36. H. Harborth, I. Mengersen. The Ramsey number of $K_{3,3}$. In *Graph Theory, Combinatorics, and Applications*, Vol. 2 (Kalamazoo, MI, 1988) (Wiley-Intersci. Publ., Wiley, New York, 1991), pp. 639–644
37. M. Chen, Y. Li, C. Pei, Gallai-Ramsey numbers of odd cycles and complete bipartite graphs. Graphs Combin. **34**(6), 1185–1196 (2018)
38. J. Zou, Y. Mao, C. Magnant, Z. Wang, C. Ye, Gallai-Ramsey numbers for books. Discrete Appl. Math. **268**, 164–177 (2019)

Chapter 4
Gallai-Ramsey Results for Other Rainbow Subgraphs

4.1 General Bounds

Recall that S_3^+ is the star on 4 vertices with the addition of an extra edge between two of the leaves. This can also be seen as a triangle with an extra pendant edge. Fujita and Magnant proved the following result.

Theorem 4.1 ([1]) *For a fixed graph H without isolated vertices, $gr_k(S_3^+ : H)$ is exponential in k if H is non-bipartite or linear if H is bipartite.*

Proof This proof is based on the proof by Fujita and Magnant [1], which follows the outline of the corresponding proof of the rainbow triangle version by Gyárfás et al. [2]. First note that the lower bounds follow directly from Theorem 3.1 since any rainbow triangle-free coloring also contains no rainbow S_3^+.

For the upper bound, first consider a general (in particular, non-bipartite) graph H. Let $t = R_3(H) - 1$ and let $n = |H|$. We claim that

$$gr_k(S_3^+ : H) \leq (2t)^{(n-1)k+1}.$$

Consider a k-coloring of a complete graph G on $(2t)^{(n-1)k+1}$ vertices. By Theorem 2.7, there exists a partition of the vertices such that there are at most 3 colors between the parts. By the choice of t, this means that there are at most t parts in this partition which implies that there exists a part H_1 with

$$|H_1| \geq \frac{|G|}{t} = 2(2t)^{(n-1)k}.$$

Let v_1 be a vertex in $G \setminus H_1$. By Theorem 2.7, the vertex v_1 has at most 2 colors on the edges to H_1. This means that v_1 has edges in a single color to at least $(2t)^{(n-1)k}$ vertices H_1' of H_1 (see Fig. 4.1). We then apply Theorem 2.7 again to find a partition of H_1' and repeat this process to find a vertex v_2.

C. Magnant and P. Salehi Nowbandegani, *Topics in Gallai-Ramsey Theory*, SpringerBriefs in Mathematics, https://doi.org/10.1007/978-3-030-48897-0_4

Fig. 4.1 The structure of G

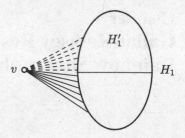

This process may be repeated $(n-1)k+1$ times since, on the last repetition, there are still $2t$ vertices remaining in G. This constructs a set of vertices $V = \{v_1, v_2, \ldots, v_{(n-1)k+1}\}$. By the pigeon hole principle and by the way, we constructed V, the graph induced on V must contain a monochromatic complete graph of order n, which contains a monochromatic copy of H. This implies that for any H, in particular if H is non-bipartite, the Gallai-Ramsey number for H is at most exponential.

We now restrict our attention to the case where H is bipartite. Let BR_3 be the bipartite Ramsey number $BR_3(H) = \min\{p : \text{any 3-coloring of } K_{p,p} \text{ contains a monochromatic copy of } H\}$. Such numbers are known to exist by a result of Erdős and Rado in [3]. Again let $t = R_3(H) - 1$ and $n = |H|$.

Given integers $m, n \geq 2$, let $z(m; n)$ denote the minimum number of edges q such that any bipartite graph $B_{m,m}$ with partite sets of size m containing at least q edges contains a copy of $K_{n,n}$. As Bollobás noted in [4], the following bound on $z(m; n)$ was proven by Znám [5].

$$z(m; n) < (n-1)^{1/n} m^{2-1/n} + \frac{n-1}{2} m = o(m^2).$$

This implies that there exists a number $b = b(n)$ which is the minimum value of m such that any bipartite graph with partite sets of sizes $m_1, m_2 \geq m$ containing at least $\frac{m_1 m_2}{2}$ edges contains a copy of $K_{n,n}$. For convenience, we let $s = (b-1)k$. Note that since H is not a star, we have $t \geq 3$. Our goal is to show that

$$gr_k(S_3^+ : H) \leq tBR_3 + s(BR_3 - 1) + b.$$

Let G be a k coloring of the complete graph on $tBR_3 + s(BR_3 - 1) + b$ vertices. By Theorem 2.7, there exists a partition of $V(G)$ with at most 3 colors on the edges between the parts. By the choice of t, this implies that there are at most t parts in this partition. Again let H_1 be a largest part of this partition and note that $|H_1| \geq BR_3$. If $|G \setminus H_1| \geq BR_3$, then there exists a monochromatic copy of H in the bipartite graph between H_1 and $G \setminus H_1$. Hence $|H_1| \geq |G| - BR_3 + 1$. Let v_1 be a vertex of $G \setminus H_1$. Note that v_1 has at most 2 colors to H_1. Associate to v_1 one of these colors which appears on at least half of the edges to H_1.

Repeat this process with H_1 in place of G to find a large part H_2 of H_1. Let v_2 be a vertex of $H_1 \setminus H_2$ and repeat to find a set $V = \{v_1, v_2, \ldots, v_{s+1}\}$. Note that

$|H_{s+1}| \geq b$. By the pigeon hole principle, there exists a set of at least b vertices $V' \subseteq V$ such that $V' \cup H_{s+1}$ contains a bipartite graph with partite sets of size b containing at least half of the edges in a single color. Since $|V'|, |H_{s+1}| \geq b$ and $|H| = n$, this bipartite graph contains a monochromatic copy of H. This proves that if H is bipartite, the Gallai-Ramsey number $gr_k(S_3^+ : H)$ is linear in k. $\square$

Recall that a broom is a star with center vertex joined to an end of a path.

Theorem 4.2 ([1]) *In any rainbow S_3^+-free coloring of a complete graph, there exists a spanning 2-colored broom.*

We include the (simple) proof by Fujita and Magnant [1].

Proof By Theorem 2.7, there exists a partition of $V(G)$ such that there are either two colors between parts or a total of 3 colors between parts with only one color between each pair of parts. If there are only two colors total, then the problem of finding a 2-colored broom reduces to the problem of finding a spanning broom in a complete multipartite graph which is trivial.

Hence, we may assume there are a total of 3 colors between parts with only one color between each pair of parts. We then arbitrarily associate two colors together to make a new coloring in which there are only 2 colors between parts and still only one between each pair of parts. At this point, we appeal to the proof of Theorem 3.3 which uses Theorem 2.1 to find a monochromatic spanning broom within the Gallai partition. This monochromatic broom corresponds to a broom on at most 2 colors in our original graph and completes the proof of Theorem 4.2. $\square$

4.2 Specific Sharp Results

In the proof of Theorem 4.3, we use Theorem 3.7 from [6].

Theorem 4.3 ([1]) *For all $t \geq 2$ and $k \geq 3$, $gr_k(S_3^+ : S_t) = 3t + 2$.*

Notice that, as in Theorem 3.7, the order does not depend on the number of colors used. In fact, the sharpness examples for Theorem 3.7 and for Theorem 4.3 use exactly 3 colors regardless of the value of k. We include the proof by Fujita and Magnant [1] with added detail.

Proof For the lower bound, consider $G = G_1 \cup G_2 \cup G_3 \cup v$ with $|G_i| = t$ where G_i is a complete graph colored entirely with color i and all edges from G_i to G_j have color ℓ where $\ell \in \{1, 2, 3\} \setminus \{i, j\}$. Finally, all edges from v to G_i are in color i (see Fig. 4.2). Every vertex of G has degree at most t in each color and since we have used only three colors, G certainly contains no rainbow S_3^+. This is a coloring of K_{3t+1} with no rainbow S_3^+ or monochromatic S_t.

For the upper bound, let G be a k coloring of the complete graph K_n where $n \geq 3t + 2$ containing no rainbow S_3^+. We break the proof into two cases based on

Fig. 4.2 The structure of G

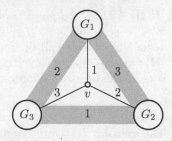

the minimum color degree of G. If $\delta^c(G) \geq 4$, then any rainbow triangle in G would imply the existence of a rainbow S_3^+. Hence, we may assume there is no rainbow triangle. By Theorem 3.7, there exists a monochromatic copy of S_t.

Finally suppose $\delta^c(G) \leq 3$ and let v be a vertex of color degree at most 3. Since v is incident to at most three colors, v is the center of a monochromatic star of size at least $\lceil \frac{n-1}{3} \rceil \geq \lceil \frac{3t+1}{3} \rceil = t + 1$. $\qquad\square$

Theorem 4.4 ([1]) *For $k \geq 1$,*

$$gr_k(S_3^+ : P_4) = k + 3.$$

We include the proof by Fujita and Magnant [1] with additional details.

Proof For completeness, we present the sharpness example from Faudree et al. [7]. Also note that this lower bound follows from Theorem 3.9. Consider a complete graph $G_1 = K_3$ colored entirely in color 1. The graph G_1 certainly contains no monochromatic copy of P_4 or rainbow copy of K_3 so certainly no rainbow copy of S_3^+. For induction, suppose we have constructed a colored complete graph G_{k-1} on $k + 1$ vertices using $k - 1$ colors with no monochromatic copy of P_4 or rainbow copy of K_3. To this, we add one vertex v with all edges out of v having color k to create a colored complete graph G_k. Note that G_k has order $k + 2$, uses k colors and contains no monochromatic copy of P_4 or rainbow copy of K_3 and so no rainbow copy of S_3^+.

For the upper bound, let G be a k-coloring of the complete graph K_n for $n = k + 3$. If $k = 1$, the result is trivial. If $k = 2$, we note that $R_2(P_4) = 5$ so any 2-coloring of K_5 yields a monochromatic copy of P_4.

If $k = 3$, label the vertices of K_6 with $a_1, \ldots, a_6$. Each vertex must have at least two edges in some color so suppose a_1 has color 1 to a_2 and a_3. This means that a_2 must have no edges of color 1 to a_i for $i \geq 4$. Hence a_2 must have two edges of a single color (not color 1) to a_i for $i \geq 4$. Suppose a_2 has edges of color 2 to a_4 and a_5. The vertices a_4 and a_5 cannot have colors 1 or 2 to a_3 so the edges a_3a_4 and a_3a_5 must have color 3. Now the vertices a_4 and a_5 cannot have any edges to a_1 or a_6 in colors 2 or 3. Hence, all edges between $\{a_4, a_5\}$ and $\{a_1, a_6\}$ must have color 1 which forces a monochromatic P_4. In fact, this implies that $R_3(P_4) = 6$. Therefore, we suppose $k \geq 4$.

By Theorem 2.7, there exists a partition of $V(G)$ such that either there are at most two colors appearing on edges between the parts, or there are at most three colors appearing on edges between the parts and all edges between each pair of parts have a single color.

First we suppose the partition results in only one color between each pair of parts and a total of 3 colors between parts. Because $R_3(P_4) = 6$, there exist at most 5 parts in the partition so since $k \geq 4$ (meaning $n \geq 7$), there exists a part A with $|A| \geq 2$.

If there exist two parts A and B each of order at least 2, then there is a copy of P_4 on the edges between A and B since all these edges have a single color. This means that at most one part A has order greater than 1. Furthermore, this means that there can be no more than 3 parts of order 1 as otherwise there would be two each having the same color on edges to A. Since $k \geq 4$, we have $|A| \geq 4$.

Let $a_1, \ldots, a_j$ be the singleton parts for $1 \leq j \leq 3$ such that a_i has all edges of color i to A (note that at least one such singleton part exists since Theorem 2.7 yields a nontrivial partition). There cannot be any edges of color i for $i \leq j$ in A as this would result in a P_4 in color i. Since $k \geq 4$, we may remove a_i for all $i \leq j$ thereby removing all edges of colors i from the graph. This produces a coloring of K_{n-j} with $k - j$ colors, no rainbow triangle and no monochromatic copy of P_4. Applying induction on k within $G \setminus \{a_1, \ldots, a_j\}$ yields a monochromatic copy of P_4, a contradiction.

Next, suppose there exist at most two colors between the parts of our partition. Since any 2-coloring of $K_{3,3}$ contains a monochromatic copy of P_4, we may assume that there is at most one part of order at least 3. In fact, this implies that if there exists a part A with $|A| \geq 3$, then there are at most 2 vertices in $G \setminus A$. Also since $R_2(P_4) = 5$, there are at most 4 parts in this partition.

Now, suppose there does not exist a part A with $|A| \geq 3$. Since each part has order at most 2, the order of the graph is $n \geq 7$ and there are at most 4 parts. Choose two parts A_1 and A_2 with order 2. Observe that all edges from $\{A_1, A_2\}$ to $G \setminus \{A_1, A_2\}$ must have at most two colors, but this is a bipartition of the graph where each part has at least 3 vertices, which is a contradiction. Hence we may assume there exists a part A with $|A| \geq 3$. Since $n \geq 7$ and $|G \setminus A| \leq 2$, we know $|A| \geq 5$. We first state a fact about the edges of a single color between A and $G \setminus A$.

Fact 4.5 *Given a vertex $v \in G \setminus A$, if the edges between v and A use more than a single color, then one color must have exactly one edge.*

Proof For a contradiction, suppose a vertex $v \in G \setminus A$ has at least two edges in each of colors 1 and 2 to A. Let $a_i, b_i \in A$ be vertices such that the edges va_i and vb_i have color i for $i \in \{1, 2\}$. In order to avoid a monochromatic copy of P_4, the edge a_1a_2 must have a third color, say color 3, for instance, (see Fig. 4.3). This means that a_1a_2v forms a rainbow triangle. In order to avoid a rainbow copy of S_3^+ or a monochromatic copy of P_4, the edges a_1b_2 and a_2b_1 must also have color 3. Hence, there is a monochromatic copy of P_4 on $b_2a_1a_2b_1$ in color 3. $\square$

First, suppose there exists only one vertex $v \in G \setminus A$. By Fact 4.5, v has at least $|A| - 1$ edges in one color (suppose color 1) to A. In order to avoid a monochromatic

Fig. 4.3 The structure of v
and A

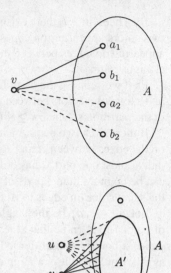

Fig. 4.4 The structure of
u, v and A

copy of P_4, there cannot be any edges of color 1 in A. Hence, we may remove v and apply induction on k as above.

Finally, suppose there exist two vertices $u, v \in G \setminus A$. Again using Fact 4.5, u and v have at least $|A| - 1$ edges of a single color to A. Suppose v has $|A| - 1$ edges to A of color 1. Let A' denote the set of vertices $a \in A$ with va in color 1. In order to avoid a monochromatic copy of P_4, the edge ua must have a different color (hence color 2) for all $a \in A'$. See Fig. 4.4. Again, since we must avoid a monochromatic copy of P_4, there must not be any edges of colors 1 or 2 in A. Hence, we may remove u and v and again apply induction on k within $G \setminus \{u, v\}$. This completes the proof of Theorem 4.4. $\qquad\square$

Theorem 4.6 ([1]) *For all $k \geq 4$, if k is even, let $\lambda(k) = 5^{k/2} + 1$ and if k is odd, let $\lambda(k) = 2 \cdot 5^{(k-1)/2} + 1$. Then*

$$gr_k(S_3^+ : K_3) = \lambda(k).$$

We include an adaptation of the proof by Fujita and Magnant [1].

Proof The proof of this result is broken into two cases. In each case, the proof involves two parts. First we show that the underlying graph in question is in fact rainbow triangle-free. Second, we apply Theorem 2.1.

For convenience, define $\lambda(3) = R_3(K_3) = 17$, $\lambda(2) = R_2(K_3) = 6$ and $\lambda(1) = R_1(K_3) = 3$. Let G be a rainbow S_3^+-free coloring of a complete graph of order $\lambda(k)$. If $\delta^c(G) \geq 4$, then, since the G is rainbow S_3^+-free, there exists no rainbow triangle in G and the desired result follows from Theorem 3.12.

If $\delta^c(G) = 1$, then let v be a vertex of color degree 1 (in color 1). Then $G \setminus v$ contains no edges of color 1 to avoid a monochromatic triangle. Also $G \setminus v$ contains no rainbow triangle T since $T \cup v$ would form a rainbow S_3^+. Hence Theorem 3.12 applies easily.

If $\delta^c(G) = 2$, let v be a vertex of degree 2 (colors 1 and 2). Let H_1 be the set of vertices adjacent to v in color 1 and let H_2 be the set of vertices adjacent to v in color 2 and, by symmetry, suppose $|H_1| \geq |H_2|$. Note that H_i contains no edge of color i and also no rainbow triangle. Hence if k is even, then $|H_1| \geq \frac{\lambda(k)-1}{2} \geq \lambda(k-1)$ so, by Theorem 3.12, H_1 contains a monochromatic triangle. If k is odd, then we know $|H_1| = |H_2| = 5^{(k-1)/2}$ and $k \geq 5$.

If both H_1 and H_2 have minimum color degree at least 3, then, aside from v, the whole graph has minimum degree 4. Since G is rainbow S_3^+-free, there must be no rainbow triangle in G and we may apply Theorem 3.12 directly. Hence, we suppose H_1 contains a vertex v' with color degree (within H_1) equal to 2.

Since v' has no edges within H_1 of color 1, suppose the two colors on edges incident to v' are 2 and 3. Let $H_{1,2}$ and $H_{1,3}$ be the subgraphs of H_1 such that v' has all edges of color 2 to $H_{1,2}$ and all edges of color 3 to $H_{1,3}$ and suppose, without loss of generality, that $|H_{1,2}| \geq |H_{1,3}|$. Since $|H_{1,2}| \geq \frac{|G|-3}{4}$ and $H_{1,2}$ contains no edges of colors 1 or 2, we may apply Theorem 3.12 to $H_{1,2}$ to complete the proof. Certainly if $d_{H_1}^c(v') = 1$, the same argument holds.

Hence, we suppose $\delta^c(G) = 3$. We divide the proof into two cases.

Case 1 *There exist no 3 monochromatic spanning trees of G with distinct colors.*

In this case, by Theorem 2.7, we may assume that G can be partitioned into two non-empty parts A and $G \setminus A$ such that the color of each edge between A and $G \setminus A$ has either color 1 or color 2. Let A be chosen so that $|A|$ is minimized. If the color of each edge between A and $G \setminus A$ is same, say, color 1, then $G \setminus A$ does not contain an edge with color 1 and also $G \setminus A$ does not contain a rainbow triangle. Applying Theorem 3.12 to $G \setminus A$, we get a contradiction. Thus we may assume that both colors 1 and 2 are used between the parts. More specifically, we claim that there is a vertex in $G \setminus A$ that has both of these colors on its incident edges to A.

Claim 4.7 *There exists a vertex $v \in G \setminus A$ such that v has at least one edge of color 1 and at least one edge of color 2 to A.*

Proof Suppose there exists no such vertex. Then, since both colors appear, there is a vertex $u \in A$ such that u is adjacent to $G \setminus A$ with both color 1 and 2 edges. Then $G \setminus A$ can be partitioned into two parts H_1 and H_2 such that every edge between A and H_i has color i. Note that H_i contains no edges of color i. Since we assume there is no vertex v as in the statement, every vertex of $G \setminus A$ must have edges of only one color (color 1 or 2) to A (see Fig. 4.5). Since G contains no monochromatic triangle, A contains no edges of color 1 or color 2. Also, since there is no rainbow triangle in A, H_1 and H_2 (otherwise, we can find a rainbow S_3^+), by Theorem 3.12, we have $|A| \leq \lambda(k-2) - 1$ and $|H_i| \leq \lambda(k-1) - 1$.

Fig. 4.5 The structure of A, H_1 and H_2

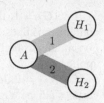

If there exists an edge e of color at least 3 (suppose color 3) between H_1 and H_2, then every vertex of A is on a rainbow triangle using colors 1, 2 and 3, (otherwise, one of the end vertices of e will be the desired vertex v). Since A contains no edges of colors 1 or 2, all edges of A must have color 3 to avoid a rainbow copy of S_3^+, so $|A| \leq 2$ to avoid creating a monochromatic triangle. This implies that $|G| \leq |H_1| + |H_2| + 2 \leq 2(\lambda(k-1) - 1) + 2 < \lambda(k)$, a contradiction. Hence, all edges between H_1 and H_2 must have either color 1 or color 2.

Suppose, for a moment, that H_1 contains no edges of color 2. Then $|H_1| \leq \lambda(k-2) - 1$ and $|H_2| \leq \lambda(k-1) - 1$ and since $G = A \cup H_1 \cup H_2$, we see that $|G| < \lambda(k)$, a contradiction. Hence we may assume that H_1 contains an edge v_1v_2 of color 2 and symmetrically that H_2 contains an edge w_1w_2 of color 1.

We have already observed that all edges between H_1 and H_2 must have color 1 or 2. For each vertex in H_1, there exists an edge of color 2 to at least one of w_1 or w_2. This provides a partition of H_1 into two (not necessarily unique) parts $H_{1,1}$ and $H_{1,2}$ such that all edges from w_i to $H_{1,i}$ have color 2. Similarly H_2 can be partitioned into $H_{2,1}$ and $H_{2,2}$ based on the edges from v_1 and v_2. Now we note that $H_{i,j}$ contains no edges of colors 1 or 2. Hence, we get $|G| \leq 5(\lambda(k-2) - 1) < \lambda(k)$ since $k \geq 4$, a contradiction. □

Take such a vertex $v \in G \setminus A$ as in Claim 4.7. Now A is partitioned into two parts A_1, A_2 such that every edge between A_i and v has color i for $i = 1, 2$. The following two claims provide additional structure. This first claim restricts the colors available for use on a short path.

Claim 4.8 *Let $P = xyz$ be a path in A. If $c(xy) = 1$, then $c(yz) \neq 2$.*

Proof Suppose that there exists a path $P = xyz$ such that $c(xy) = 1$ and $c(yz) = 2$. By the partition of G, $G \setminus A$ can be partitioned into two parts H_1 and H_2 such that every edge between H_i and y has color i for $i = 1, 2$. Since there is no monochromatic triangle in G, every edge between x and H_1 has color 2 and every edge between z and H_2 has color 1. This means that each H_i does not contain any edges of colors 1 or 2. Then applying Theorem 3.12 to the larger of H_1 or H_2, we get a contradiction. □

This next claim provides an edge of color 4 within A.

Claim 4.9 *There exists an edge of color at least 4 in A.*

Proof Suppose not, then A contains at most 3 colors. By Claim 4.8, the edges of color 1 and color 2 in A do not share a vertex. Consider the (disjoint) subgraphs of $G_i \cap A$ for $i = 1, 2$. Since A contains at most 3 colors, all edges between these subgraphs must have color 3. This means that the vertices of $G_1 \cap A$ and $G_2 \cap A$ must not contain any edges of color 3 and if both subgraphs are non-empty, then $|A| = 4$. Otherwise, A has only two colors on the edges so, in either case $|A| \leq 5$.

Next we claim that A does not contain any edges of color 1 or 2 (hence, A is a single edge of color 3). Suppose that A contains an edge xy of color 1 or 2 (suppose color 1). Bipartition $G \setminus A$ into sets H_i such that all edges from x to H_i have color i for $i = 1, 2$. In order to avoid a monochromatic triangle in color 1, all edges from y to H_1 must have color 2. By Theorem 3.12, this means that $|H_1| \leq \lambda(k - 2) - 1$ and $|H_2| \leq \lambda(k - 1) - 1$. This, along with the fact that $|A| \leq 5$ contradicts the choice of G with $|G| \geq \lambda(k)$ unless $k = 4$.

If $k = 4$, the above counting implies that $|A| = 5$, $|H_1| = 5$ and the subgraph induced on A consists of two 5-cycles (in colors 1 and 3) and the subgraph induced on H_1 consists of two 5-cycles (in colors 3 and 4). Label the vertices of the 5-cycle of color 1 in A with $x_1, \ldots, x_5$ with $x = x_1$. This means that x_2 and x_5 must have all edges of color 2 to H_1. If x_3 (similarly x_4) had an edge of color 2 to H_1, this, along with the edge of color 3 to x_1 and an edge of color 4 in H_1 would form a rainbow S_3^+. Hence, x_3 and x_4 must have all edges of color 1 to H_1, which forms a triangle in color 1. This means that A is a single edge of color 3, which contradicts the assumption that $|A| = 5$. This proves that A contains no edge of color 1 or 2, meaning that A is a single edge of color 3.

By the choice of the vertex v, we know that v and the two vertices of A form a rainbow triangle with color 1, 2, 3. Let H_i' be the tripartition of $G \setminus (A \cup v)$ such that each edge between v and H_i' has color i for $i = 1, 2, 3$. Let a_i be the vertex of A with an edge of color i to v for $i = 1, 2$. In order to avoid a monochromatic triangle, all edges from a_1 to H_1' must have color 2 and similarly all edges from a_2 to H_2' must have color 1. This implies that $|H_i'| < \lambda(k - 2)$ for $i \leq 2$. If we bipartition H_3' into $H_{3,1}'$ and $H_{3,2}'$ such that a_1 has all edges of color i to $H_{3,i}'$, we also find that $|H_{3,i}'| < \lambda(k - 2)$. This implies that $|G| \leq 4\lambda(k - 2) < \lambda(k)$, a contradiction. $\square$

Since there is no rainbow S_3^+, the color of every edge between A_1 and A_2 is one of $\{1, 2, 3\}$. Hence, without loss of generality, we may suppose that A_1 contains an edge of color 4. Let X be a maximal component of $G_{\geq 4}$ in A_1. Note that the color of every edge between X and A_2 must be 1 or 2 because G does not contain a rainbow copy of S_3^+. Then, by the minimality of $|A|$, we obtain the fact below.

Fact 4.10 *The set $X \neq A_1$, and furthermore, there exists a vertex $y \in A_1 \setminus X$ with an edge of color 3 to A_2.*

We now prove some additional claims which allow us to complete the proof of this case. This next claim provides some additional structure about A_2.

Claim 4.11 *If* $|A_2| = 1$, *let* $\{z\} = A_2$, *then one of the following holds:*

1. *There exist at least two independent edges in A of color 2.*
2. $|A_1| \leq |X| + 2$ *and every edge between X and z has color i (for some $i \leq 2$) and every edge between X and y has color 3.*

Proof Suppose there are not two independent edges in A of color 2. Recall that z must have all edges in a single color i to X for some $i \leq 2$. By Claim 4.8, the vertices of X have no other edges of color i. Since there are no two independent edges in color 2, we see that there is exactly one nontrivial component (being X) of $G_{\geq 4}$ in A_1 and by the minimality of $|A|$, we see that y must have all edges of color 3 to X. There may, however, be at most one vertex y' in $A_1 \setminus (X \cup y)$ since more vertices would force either two disjoint edges in color 2 or a monochromatic triangle in color 3. This is precisely the claimed structure, completing the proof of the claim. $\square$

Next we show that A_2 is not too small.

Claim 4.12 $|A_2| \geq 2$.

Proof Suppose, for a contradiction, that $|A_2| = 1$, say with $A_2 = \{z\}$. Note that there exists no vertex $v' \in G \setminus A$ which has at least two edges of each color 1 and 2 to A (otherwise we would have chosen v' as opposed to v). If Case 1 of Claim 4.11 holds, then we let Y be the set of vertices in $G \setminus A$ which have at least $|A| - 1$ edges of color 2 to A. If Case 2 of Claim 4.11 holds, we define Y to be the set of vertices with at least $|A| - 1$ edges of color i to A (where i is determined in Claim 4.11). Suppose that $i = 2$ (we will consider the case when $i = 1$ later). We now show that $|Y|$ is small.

Certainly no vertex in $G \setminus A$ has only edges of color 2 to A since this would form a monochromatic triangle (by Claim 4.11). This means that every vertex of Y has exactly $|A| - 1$ edges of color 2 to A and one edge of color 1 to A. This implies immediately that Y contains no edges of color 2. Since every vertex of Y is on a rainbow triangle with colors $1, 2, 3$ (by Fact 4.10), we see that Y contains only edges of colors 1 and 3, meaning that $|Y| \leq 5$. By our assumptions, every vertex of $(G \setminus A) \setminus Y$ must have at least $|A| - 1$ edges of color 1 to A. Hence, $|A| \leq (\lambda(k - 2) - 1) + 3$ since there exists a set D of at most three vertices in A such that every edge of color 1 or 2 in A contains at least one vertex of D (i.e. the removal of the set D would destroy all edges of color 1 and 2 in A).

If there exists no edge of color 1 in $G \setminus A$, then since $|A| \leq |G \setminus A|$, we have

$$|G| = |A| + |G \setminus A| \leq 2(\lambda(k - 1) - 1) < \lambda(k),$$

which is a contradiction. Hence, let e be an edge of color 1 in $G \setminus A$. Since every vertex of $(G \setminus A) \setminus Y$ has many edges of color 1 to A, at least one end vertex of the edge e must be in the set Y in order to avoid a triangle in color 1. This implies that there are at most $|Y| \leq 5$ disjoint edges of color 1 in $G \setminus A$ (i.e. the removal of Y would destroy all edges of color 1 in $G \setminus A$). Hence

$$|G| = |A| + |G \setminus A|$$
$$\leq [\lambda(k-2) + 2] + [(\lambda(k-1) - 1) + 5]$$
$$= \lambda(k-2) + \lambda(k-1) + 6$$
$$< \lambda(k)$$

since $k \geq 4$. This is a contradiction, completing the proof in this case.

Now we suppose that Case 2 of Claim 4.11 holds and that $i = 1$. Note that X contains no edges of colors 1 or 3 so $|A| \leq |X| + 3 \leq \lambda(k-2) + 2$. By Claim 4.10, every vertex of Y is in a rainbow triangle (with two vertices of A) so Y contains only edges of colors 1, 2 or c', for some color $c' \geq 3$. Since every vertex of Y has many edges of color 1 to A, we see that Y contains no edges of color 1. This implies that $|Y| \leq 5$. As above, we can easily see that every edge of color 2 in $G \setminus A$ must have at least one end in Y. This means that $|G \setminus A| \leq \lambda(k-1) + 4$ and again we arrive at the same contradiction as above. □

Let $v^* \in A_1 \setminus X$ and suppose v^* is adjacent to X with an edge of color 2, say, $c(v^*z) = 2$ for some $z \in X$. In view of Claim 4.8, we see that there is no edge with color 2 between v^* and A_2 since otherwise we can form either a rainbow copy of S_3^+ or a monochromatic triangle with color 2 by using vertices v^* and z. Again by Claim 4.8, it follows that every edge between v^* and A_2 has color 3. Then, since z is not contained in any rainbow triangle, every edge between A_2 and z has color 2. Applying Claim 4.8 to an edge of color 2 between A_2 and z, we conclude that every edge between A_2 and X has color 2. Again by Claim 4.8, A_2 does not contain color 1. Also, since there is no monochromatic triangle, A_2 does not contain an edge of color 2 or 3. Since $|A_2| \geq 2$, A_2 has an edge ab with the color distinct from 1, 2, 3. Then we can find a rainbow copy of S_3^+ using vv^*ab because every edge between v^* and A_2 has color 3, a contradiction.

Hence, every edge between $A_1 \setminus X$ and X has color 3. By Fact 4.10, there exists a pair of vertices $y \in A_1 \setminus X$ and $y' \in A_2$ such that $c(yy') = 3$. Suppose that there exists a vertex $z' \in A_2$ such that $c(y'z') = 1$. Then by Claim 4.8, every edge between z' and X has color 1, and again in view of Claim 4.8, every edge between A_2 and X has color 1. However, then we can find a monochromatic K_3 with color 1 by using $y'z'$, a contradiction. Hence, since $\{v, y, y'\}$ forms a rainbow triangle, every edge within A_2 which contains y' has color 3. Since $|A_2| \geq 2$, take $x' \in A_2 \setminus y'$. Note that $c(x'y') = 3$. Since there is no monochromatic copy of K_3 with color 3, we see that $c(x'y) \in \{1, 2\}$. Then by Claim 4.8, every edge between A_2 and X has color $c(x'y)(= 1$ or 2).

If $c(x'y) = 1$ (an identical process holds if $c(x'y) = 2$), we partition $G \setminus A$ into H_1 and H_2 such that $c(yu) = i$ for $u \in H_i$. In order to avoid a monochromatic triangle (using y and a vertex of H_1), x' must have all edges of color 2 to H_1. Hence, H_2 uses $k - 1$ colors, and H_1 and A_2 use only $k - 2$ colors.

If there exists a vertex $y^* \in A_1 \setminus X \cup \{y\}$, then y^* has only edges of color 3 to X and y which is clearly a contradiction. Hence, $|A| = |X| + 1$ and X uses only $k - 3$ colors. If there exists no edge of color at least 4 in A_2, then A_2 uses only 2 colors so $|A_2| < r_2(K_3) = 6$. If there are edges of color at least 4 in A_2, then by symmetry, we get the same structure as in A_1. Just like in A_1, all but 1 vertex of A_2 use at most $k - 3$ colors.

Hence, if $k \geq 5$, we have $|H_1| \leq \lambda(k - 2) - 1$, $|H_2| \leq \lambda(k - 1) - 1$, and $|A_1|$, $|A_2| \leq \lambda(k - 3)$. Since $G = A_1 \cup A_2 \cup H_1 \cup H_2$, for $k \geq 5$, we get

$$\begin{aligned}
|G| &= |A_1| + |A_2| + |H_1| + |H_2| \\
&\leq 2\lambda(k - 3) + \lambda(k - 2) + \lambda(k - 1) - 2 \\
&< \lambda(k).
\end{aligned}$$

This contradicts the original assumption that $|G| \geq \lambda(k)$. If $k = 4$, we may have $|A_2| = 5 > 2 = \lambda(k - 3)$ but an identical count still provides the desired result, completing the proof of the first case.

Case 2 *There are 3 monochromatic spanning trees with distinct colors.*

Assume that the 3 monochromatic spanning trees have colors say $1, 2, 3$, respectively. Let v be a vertex with $d^c(v) = 3$, and partition $G \setminus v$ (by Theorem 2.7) into sets A_i such that every edge between v and A_i has color i for $i = 1, 2, 3$. Hence $G = A_1 \cup A_2 \cup A_3 \cup \{v\}$. We prove the following key claim.

Claim 4.13 *Suppose that A_i contains an edge with color $j \geq 4$. Then for each A_s with $s \neq i$, $|A_s| \leq 3(\lambda(k - 3) - 1) + 1$.*

Proof In this proof, by symmetry, we may assume that $i = 1$, $j = 4$ and $s = 2$. Take an edge e with color 4 in A_1. In order to avoid creating a rainbow copy of S_3^+, it is easy to see that A_2 can be partitioned into two parts B_1, B_2 such that every edge between B_ℓ and e has color ℓ for $\ell = 1, 2$ (note that it is possible to have $B_\ell = \emptyset$ for some ℓ).

Since G does not contain a monochromatic copy of K_3 with color 2 or a rainbow copy of S_3^+, it follows that every edge between B_1 and B_2 has color 1. This means that $|B_\ell| \leq \lambda(k - 2) - 1$ since B_ℓ contains no edges of colors 1 or 2 for $\ell = 1, 2$. If there is no edge with color at least 4 in A_2, then $|B_\ell| \leq \lambda(k - 3) - 1$ and so $|A_2| \leq 2(\lambda(k - 3) - 1)$, which is the desired result. So, suppose that B_1 contains an edge with color at least 4 (an identical argument holds for B_2).

Let X be a component (containing at least one edge) of $G_{\geq 4}$ in B_1. By the assumption of Case 2, there exist vertices x, y such that $x \in X$ and $y \in (A_3 \cup B_1) \setminus X$ and $c(xy) = 3$. Since G does not contain a rainbow copy of S_3^+, every edge between X and y has color 3. This means that the set X does not contain an edge with color 3. Hence, $|X| \leq \lambda(k - 3) - 1$. Also, note that all edges between components of $G_{\geq 4}$ in B_1 must have color 3, which implies that there are at most two such components and $|B_1| \leq 2(\lambda(k - 3) - 1)$.

By the assumption of Case 2, there exists an edge of color 3 from a vertex $x' \in B_1$ to a vertex $y' \in A_3$. Since x' is adjacent to all of B_2 with edges of color 1 and G contains no rainbow copy of S_3^+, every nontrivial (more than a single vertex) component of $G_{\geq 4}$ in B_2 must have all edges of color 3 to y'. This means there can be at most one such component in B'. By the same argument as above, this means that $|B_2| \leq (\lambda(k-3)-1)+1$. Together, this implies that $|A_2| \leq 3(\lambda(k-3)-1)+1$. $\qquad\square$

First suppose that more than one set A_i contains edges of color at least 4. Then, by Claim 4.13, $|A_i| \leq 3(\lambda(k-3)-1)+1$ and $|G| \leq 9(\lambda(k-3)-1)+4 < \lambda(k)$ which is a contradiction. Hence, suppose only A_1 contains edges of color at least 4. This implies that $|A_2|, |A_3| \leq 5$ since each set contains only two colors. Since $k \geq 4$, we see that $|G| \leq |A_1| + |A_2| + |A_3| + 1 \leq [\lambda(k-1)-1] + 11 < \lambda(k)$ if $k \geq 5$, another contradiction. If $k = 4$, Claim 4.13 implies that $|A_2|, |A_3| \leq 2$ which means $|G| \leq [\lambda(k-1)-1]+5 < \lambda(k)$, yet another contradiction. This completes the proof of Theorem 4.6. $\qquad\square$

Continuing in the theme of cycles, we provide the number for C_4, an extension of Theorem 3.10.

Theorem 4.14 ([1]) *For $k \geq 4$, $gr_k(S_3^+ : C_4) = k + 4$.*

Note that for $k \leq 3$, $gr_k(S_3^+ : C_4) = R_k(C_4)$. We include a modified version of the proof by Fujita and Magnant [1].

Proof The lower bound follows from Theorem 3.10 since any rainbow triangle-free construction is also free of a rainbow copy of S_3^+.

Let G be a rainbow S_3^+-free coloring of a complete graph K_n with $n = k + 4$. By contradiction, assume that G does not contain a monochromatic copy of C_4.

Claim 4.15 *The minimum color degree satisfies $\delta^c(G) \geq 2$.*

Proof Suppose $\delta^c(G) = 1$. Then, there exists a vertex-sequence $v_1, \ldots, v_t$ of G with $t \geq 1$ such that every edge between v_i and $G \setminus \{v_1, \ldots, v_i\}$ has a single color c_i for each i with $1 \leq i \leq t$. Choose such a vertex-sequence with t maximum. Since G does not contain a monochromatic copy of C_4, it follows that $c_i \neq c_j$ holds for $i \neq j$, and for each i with $1 \leq i \leq t$, $G \setminus \{v_1, \ldots, v_i\}$ does not contain a monochromatic copy of P_3 with color c_i. (Note that this also implies $G \setminus \{v_1, \ldots, v_j\}$ does not contain a monochromatic copy of P_3 with color c_i for all $j \geq i$.)

Let $G' = G \setminus \{v_1, \ldots, v_t\}$. Note that $|G'| \geq 4$. Since $k \geq 4$ and G contains no rainbow copy of S_3^+, we see that $t \leq k - 1$. By the maximality of t, we get $\delta^c(G') \geq 2$.

First we claim that G' contains a rainbow triangle. Suppose that G' is rainbow triangle-free. Then, by Theorem 2.1, we see that G' can be partitioned into $H_1, \ldots, H_\ell$ with $|H_1| \leq \ldots \leq |H_\ell|$ and $\ell \geq 2$ such that every edge between each pair H_i, H_j, has a single color, c_{t+1} or c_{t+2} (note that $c_{t+1} \neq c_{t+2}$ and if $t = k - 1$, c_{t+2} would not occur).

Since G does not contain a monochromatic C_4, this implies that $|H_i| = 1$ for each i with $1 \leq i \leq \ell - 1$. Also, since $\delta^c(G') \geq 2$, we see from $|H_1| = 1$ that both c_{t+1} and c_{t+2} are used in the above partition, and this implies $\ell \geq 4$. On the other hand, since $R_2(C_4) = 6$, this implies $\ell \leq 5$. Thus we have $4 \leq \ell \leq 5$.

Since $\ell \geq 4$ and G does not contain a monochromatic copy of C_4, it is easy to check that there exist two monochromatic copies of P_3 such that one of them has color c_{t+1} and the other has color c_{t+2}. This means that $t \leq k - 2$, and hence $|G'| \geq 6$. Since $\ell \leq 5$, we have $|H_\ell| \geq 2$. Then, we can find a monochromatic copy of C_4 with color c_{t+1} or c_{t+2} in $H_i \cup H_j \cup H_\ell$ for some i, j because $\ell \geq 4$. This is a contradiction.

Hence, there exists a rainbow triangle T in G'. Since $t \geq 1$ and G is rainbow S_3^+-free, this implies that T contains colors c_j for $1 \leq j \leq t$. Moreover, since T uses only 3 colors, we have $t \leq 3$. Since $k \geq 4$, there exists a fourth color c_4 appearing on at least one edge in G'.

Note that no edge between T and $G' \setminus T$ has color c_4. Hence, there exists an edge $e = xy$ with color c_4 such that $x, y \in G' \setminus T$. Let u, v, w be vertices in T such that uv has color c_j for some $j \leq t$. Also, let c_{uw}, c_{vw} be colors used in uw, vw, respectively. Consider the colors of edges between uv and xy. Since there is no monochromatic P_3 with color c_j in G', each edge between uv and xy has color c_{uw} or c_{vw} because G is rainbow S_3^+-free. Then we can easily find a monochromatic C_4 or a rainbow S_3^+ in $T \cup \{x, y\}$. This is a contradiction, completing the proof of the claim.   □

Suppose that $\delta^c(G) \geq 4$. Then, G does not contain a rainbow triangle because G is S_3^+-free, and hence there exists a Gallai partition. Since the order of the smallest component is at least three (because $\delta^c(G) \geq 4$), we can easily find a monochromatic C_4 using edges between the parts of Gallai partition, a contradiction. Note that we may also apply Theorem 3.10 in this case. Hence, by Claim 4.15, it follows that $2 \leq \delta^c(G) \leq 3$.

Suppose that there exist three monochromatic spanning trees with distinct colors in G. In this case, we have $\delta^c(G) = 3$. Then, by Theorem 2.7, we can find a partition of $V(G)$ such that there are three colors on the edges between the parts and between each pair of parts, there is only one color.

Let $H_1, \ldots, H_\ell$ be the parts of the partition with $|H_1| \leq \ldots \leq |H_\ell|$ and $\ell \geq 2$. We see that $|H_{\ell-1}| = 1$ since otherwise we can find a monochromatic copy of C_4. Since $\delta^c(G) = 3$, we see that $\ell \geq 4$. If $\ell \geq 5$, then we can find a monochromatic copy of C_4 in $H_i \cup H_j \cup H_\ell$ for some $1 \leq i < j \leq \ell - 1$ because the total number of colors used in the partition is three. Thus we have $\ell = 4$. However, in this case, we can easily find a monochromatic copy of C_4 or a rainbow copy of S_3^+ in G because the part H_ℓ must contain a fourth color. This is a contradiction. Hence we may assume that G does not contain three monochromatic spanning trees. Then, by Theorem 2.7, there exists a partition of G with at most two colors on the edges between the parts.

The proof is divided into two cases based on $\delta^c(G)$. First we note an easy fact.

Fact 4.16 *Given two disjoint edges e and f with different colors (suppose colors 1 and 2), the edges between e and f must have at least one edge of color 1 or 2.*

The proof of this fact is a simple consideration of the possible colors on the edges.

Case 1 $\delta^c(G) = 3$.

Let H be a component of the partition. By the conditions on the partition, we may assume that every edge between H and $G \setminus H$ has color c_1 or c_2 (with $c_1 \neq c_2$).

Note that since $\delta^c(G) = 3$, we have $|H| \geq 2$ and $|G \setminus H| \geq 2$. Since $\delta(G) = 3$ there exists an edge e of color c_3 in H. Furthermore, there exists an edge f which does not have color c_1 or c_2 in $G \setminus H$. Since there are only edges of colors c_1 and c_2 between e and f, by Fact 4.16 we may assume f has color c_3. Since $k \geq 4$, there must exist an edge g with color c_4 somewhere in the graph. This edge certainly does not go between H and $G \setminus H$ by the nature of the partition. Hence, without loss of generality, the edge g lies entirely within H. Then g and f form a contradiction to Fact 4.16 completing the proof in this case.

Case 2 $\delta^c(G) = 2$

Let v be a vertex in G with $d^c(v) = 2$. We may assume that v has an edge with color c_i for each $i = 1, 2$. Then $G - v$ can be partitioned into two parts A_1, A_2 such that all edges between v and A_i has color c_i for $i = 1, 2$.

First suppose that $|A_1| \geq 2$ and $|A_2| \geq 2$. Since $|G \setminus v| \geq 5$, we see that there is an edge between A_1 and A_2 with the 3rd color c_3 (otherwise we can find a monochromatic C_4 with color c_1 or c_2). Let uw be an edge which color c_3 such that $u \in A_1$, $w \in A_2$. Note that $\{u, v, w\}$ forms a rainbow triangle with colors c_1, c_2, c_3. Since $k \geq 4$, there is an edge $e = xy$ with 4th color c_4 in G. Since G is rainbow S_3^+-free, we see that x, y, u, v, w are all distinct. Suppose that $|A_1 \cap \{x, y\}| = |A_2 \cap \{x, y\}| = 1$, say, $x \in A_1$, $y \in A_2$. Then, by considering the edges between $\{x, y\}$ and $\{u, w\}$, by Fact 4.16 there is an edge in color either c_3 or c_4. This edge along with v and one of $\{x, y\}$ or $\{u, w\}$ forms a rainbow copy of S_3^+, a contradiction. Hence, by symmetry of the roles of the sets A_i, we may assume that $x, y \in A_1$. Recall that there exists a vertex $z \in A_2 \setminus w$. Again, by considering the edges between $\{x, y\}$ and $\{u, w\}$ and using the vertex z, we find either a rainbow copy of S_3^+ or a monochromatic copy of C_4. Using Fact 4.16 on xy and uw, we obtain a contradiction.

Thus, we may assume that $|A_1| = 1$. Let $A_1 = \{u\}$ and we first suppose that $d^c(u) = 2$. Since G does not contain a monochromatic copy of C_4, note that there is at most one edge with color c_2 between u and $G \setminus \{u, v\}$. This implies that at least $|G \setminus \{u, v\}| - 1$ edges between u and $G \setminus \{u, v\}$ have another color. If the color is distinct from c_1, c_2, say, it is c_3, then note that there are at least $|G \setminus \{u, v\}| - 1$ vertices each contained in a rainbow triangle with colors c_1, c_2, c_3. We may then easily find a rainbow copy of S_3^+ in G because $k \geq 4$. This is a contradiction.

Hence, we may assume that the colors of all edges used between u and $G \setminus \{u, v\}$ are both c_1 and c_2. Let z be a vertex in $G \setminus \{u, v\}$ such that zu has color c_2. Since $k \geq 4$, we may assume that there exist two edges $e = xy$ and $e' = x'y'$ in $G \setminus \{u, v\}$ with colors c_3, c_4, respectively, where $c_3, c_4 \notin \{c_1, c_2\}$. Since G is rainbow S_3^+-free, either $z \cap \{x, y\} = \emptyset$ or $z \cap \{x', y'\} = \emptyset$ holds (because u, z and another vertex distinct from v forms a rainbow triangle). By symmetry, we may assume that $z \cap \{x, y\} = \emptyset$.

Since G does not contain a monochromatic copy of C_4 the edges from z to $\{x, y\}$ may not both be color c_1 or c_2. Also, since G does not contain a rainbow S_3^+, the vertex z must not have an edge of a different color to $\{x, y\}$. Hence, $\{x, y, z\}$ forms a rainbow triangle with color c_1, c_2, c_3. This implies that $\{x, y, z\} \cap \{x', y'\} = \emptyset$. Then, by considering the edges between $\{x, y, z\}$ and $\{x', y'\}$ and using u, v, we can find either a monochromatic copy of C_4 or a rainbow copy of S_3^+, a contradiction.

Thus we have $d^c(u) \geq 3$. Assume that u is adjacent to an edge with color c_3 and let $A_3 := \{z \in A_2 | uz \text{ is } c_3\}$. Since u is contained in a rainbow triangle uvz for some $z \in A_3$, with colors c_1, c_2, c_3, note that $d^c(u) = 3$ and the colors which are adjacent to u are precisely c_1, c_2, c_3.

Let $z \in A_3$. Since $k \geq 4$, there exists an edge $e = xy$ in $G \setminus \{u, v\}$ such that e has color c_4 with $c_4 \notin \{c_1, c_2, c_3\}$. The graph G is rainbow S_3^+-free, so $z \cap \{x, y\} = \emptyset$. Since G contains no monochromatic C_4 or rainbow S_3^+, it is easy to see that both ux and uy have color c_1. Arguing similarly, we also see that both zx and zy have color c_3. Let $w \in G \setminus \{x, y, z, u, v\}$ (this is possible because $|G| \geq 6$). Then, observing edges between w and $\{x, y\}$, we can find either a rainbow S_3^+ or a monochromatic C_4. This is a contradiction, completing the proof of Case 2, and hence this completes the proof of the theorem. $\qquad\square$

References

1. S. Fujita, C. Magnant, Extensions of Gallai-Ramsey results. J. Graph Theory **70**(4), 404–426 (2012)
2. A. Gyárfás, G. Sárközy, A. Sebő, S. Selkow, Ramsey-type results for gallai colorings. J. Graph Theory **64**(3), 233–243 (2010)
3. P. Erdős, R. Rado, A partition calculus in set theory. Bull. Am. Math. Soc. **62**, 427–489 (1956)
4. B. Bollobás, *Extremal Graph Theory* (Dover Publications Inc., Mineola, NY, 2004). Reprint of the 1978 original
5. Š. Znám, On a combinatorical problem of K. Zarankiewicz. Colloq. Math. **11**, 81–84 (1963)
6. A. Gyárfás, G. Simonyi, Edge colorings of complete graphs without tricolored triangles. J. Graph Theory **46**(3), 211–216 (2004)
7. R.J. Faudree, R. Gould, M. Jacobson, C. Magnant, Ramsey numbers in rainbow triangle free colorings. Australas. J. Combin. **46**, 269–284 (2010)

Chapter 5
Conclusion and Open Problems

This chapter concludes the book by summarizing the known sharp results to the best of our knowledge with appropriate citations. We also include a section with conjectures and open problems for those interested in pursuing future work in the area. These problems are by no means exhaustive since any unknown sharp value of a Gallai-Ramsey number would contribute to the literature.

5.1 Summary Table of Known Results

In this section, we provide tables of known sharp results for Gallai-Ramsey numbers.

© The Author(s), under exclusive license to Springer Nature Switzerland AG 2020   97
C. Magnant and P. Salehi Nowbandegani, *Topics in Gallai-Ramsey Theory*,
SpringerBriefs in Mathematics, https://doi.org/10.1007/978-3-030-48897-0_5

First a table of results for several small sparse graphs.

Graph	Notation	$gr_k(K_3 : H)$	Cite
	P_4	$k+3$	[1]
	P_5	$k+4$	[1]
	P_6	$2k+4$	[1]
	P_7	$2k+5$	[2]
	P_8	$3k+5$	[2]
	P_4^+	$k+4$	[1]
	P_5^+	$k+5$	[1]
	P_4^{++}	$2k+4$	[1]
	P_4^{+2}	$2k+4$	[1]
	$P_5^{+'}$	$k+5$	[1]
	C_4	$k+4$	[1]
	C_5	$2^{k+1}+1$	[3]
	C_6	$2k+4$	[3]
	C_7	$3 \cdot 2^k + 1$	[4]
	C_8	$3k+5$	[5]
	C_9	$4 \cdot 2^k + 1$	[6]
	C_{11}	$5 \cdot 2^k + 1$	[6]
	C_{13}	$6 \cdot 2^k + 1$	[7]
	C_{15}	$7 \cdot 2^k + 1\cdot$	[7]

Next a table of results for several small dense graphs.

Graph	Notation	$gr_k(K_3 : H)$	Cite
	K_3	$\begin{cases} 5^{k/2} + 1 & k \text{ even}, \\ 2 \cdot 5^{(k-1)/2} + 1 & k \text{ odd} \end{cases}$	[8–10]
	K_4	$\begin{cases} 17^{k/2} + 1 & k \text{ even}, \\ 3 \cdot 17^{(k-1)/2} + 1 & k \text{ odd} \end{cases}$	[11]
	K_5	See Theorem 3.21	[12]
	$K_{3,3}$	$2k + 14$	[13]
	$sK_3, (k-s)C_4$	$\begin{cases} (k-s+3) \cdot 2 \cdot 5^{(s-1)/2} + 1 & \text{if } s \text{ is odd and } s < k - 1, \\ (k-s+3) \cdot 5^{s/2} + 1 & \text{if } s \text{ is even and } s < k - 1, \\ 6 \cdot 5^{(s-1)/2} + 1 & \text{if } s \text{ is odd and } s = k - 1, \\ 3 \cdot 5^{s/2} + 1 & \text{if } s \text{ is even and } s = k - 1, \\ 2 \cdot 5^{(s-1)/2} + 1 & \text{if } s \text{ is odd and } s = k, \\ 5^{s/2} + 1 & \text{if } s \text{ is even and } s = k \end{cases}$	[14]
	B_2	$\begin{cases} 16 \cdot 5^{(k-2)/2} + 1 - \ell & \text{if } k \text{ is even}, \\ 32 \cdot 5^{(k-3)/2} + 1 - \ell & \text{if } k \geq 3 \text{ is odd} \end{cases}$	[15]
	B_3	$\begin{cases} 22 \cdot 5^{(k-2)/2} + 1 - 2\ell & \text{if } k \text{ is even}, \\ 44 \cdot 5^{(k-3)/2} + 1 - 2\ell & \text{if } k \geq 3 \text{ is odd} \end{cases}$	[15]
	B_4	$\begin{cases} 28 \cdot 5^{(k-2)/2} + 1 - 3\ell & \text{if } k \text{ is even}, \\ 56 \cdot 5^{(k-3)/2} + 1 - 3\ell & \text{if } k \geq 3 \text{ is odd} \end{cases}$	[15]
	B_5	$\begin{cases} 32 \cdot 5^{(k-2)/2} + 1 - 4\ell & \text{if } k \text{ is even}, \\ 64 \cdot 5^{(k-3)/2} + 1 - 4\ell & \text{if } k \geq 3 \text{ is odd} \end{cases}$	[15]

Finally a table of sharp results for some classes of graphs.

Graph	Notation	$gr_k(K_3 : H)$	Cite
	$K_{1,m}$	$\begin{cases} \frac{5m-6}{2} & \text{if } m \text{ is even}, \\ \frac{5m-3}{2} & \text{if } m \text{ is odd} \end{cases}$	[16]
	$K_{1,m}^+$	$\begin{cases} (2m)^{k/2} + 1 & \text{if } k \text{ is even}, \\ m \cdot (2m)^{(k-1)/2} + 1 & \text{if } k \text{ is odd} \end{cases}$	[17]
	$C_{2\ell+1}$	$\ell \cdot 2^k + 1$	[18, 19]

5.2   Open Problems

For bipartite graphs, we believe the following to be true. Recall for a connected bipartite graph H, we have defined $s(H)$ to be the order of the smallest part in the unique bipartition of H.

Conjecture 3 ([13]) For $k \geq 3$, if H is connected and bipartite, then

$$gr_k(K_3 : H) = R_2(H) + (k - 2)(s(H) - 1).$$

The lower bound of Conjecture 3 follows from Proposition 3.1. Conjecture 3 holds true for all known values of $gr_k(K_3 : H)$ for which H is bipartite.

More specifically for even cycles, we believe the following to be true.

Conjecture 4 For all $\ell \geq 3$ and $k \geq 2$, we have

$$gr_k(K_3 : C_{2\ell}) = (\ell - 1)(k + 1) + 2.$$

For books, the sharp value of the Gallai-Ramsey number was conjectured in [15].

Conjecture 5 ([15]) If B_m is the book with m pages, $B_m = K_2 + \overline{K_m}$, then for $k \geq 2$,

$$gr_k(K_3 : B_m) = \begin{cases} (R(B_m, B_m) - 1) \cdot 5^{(k-2)/2} + 1 & \text{if } k \text{ is even,} \\ 2 \cdot (R(B_m, B_m) - 1) \cdot 5^{(k-3)/2} + 1 & \text{if } k \text{ is odd.} \end{cases}$$

For complete graphs, the value of the Gallai-Ramsey number was conjectured in [20] (also stated as Conjecture 1).

Conjecture 6 ([20]) For $k \geq 1$ and $p \geq 3$,

$$gr_k(K_3 : K_p) = \begin{cases} (r(p) - 1)^{k/2} + 1 & \text{if } k \text{ is even,} \\ (p - 1)(r(p) - 1)^{(k-1)/2} + 1 & \text{if } k \text{ is odd.} \end{cases}$$

Conjecture 1 is known to hold for $p \leq 4$ by Theorems 3.12 and 3.19.

More generally, for any non-bipartite graph, we believe the following to be true. Recall the definition of $m(H)$ in Definition 3.1 from [21].

Conjecture 7 ([21]) For a connected non-bipartite graph H and an integer $k \geq 2$, we have that $gr_k(K_3 : H)$ is at least

$$\begin{cases} (R(H, H) - 1) \cdot (m(H) - 1)^{(k-2)/2} + 1 & \text{if } k \text{ is even,} \\ (\chi(H) - 1) \cdot (R(H, H) - 1) \cdot (m(H) - 1)^{(k-3)/2} + 1 & \text{if } k \text{ is odd.} \end{cases}$$

These very ambitious conjectures are not likely to be completely solved any time soon but provide a healthy bank of problems to pursue.

Recall the precise definition of $gr_k(G : H)$, particularly with the line about using "at most k colors". Similarly, also define $gr'_k(H : G)$ to be the minimum order of K_n in which any edge-coloring using *precisely all k colors* contains either a rainbow colored copy of H or a monochromatic copy of G. Certainly $gr'_k(H : G) \leq gr_k(H : G)$ always holds. Note that if $gr_k(H : G)$ is a monotone increasing function of k (on an interval $a \leq k \leq b$ with $a < b$), then these two functions will be equal (on the same interval). Somewhat surprisingly, this is not always the case. Recall Theorem 4.3 for the value of $gr_k(S_3^+ : S_t)$.

Regarding $gr'_k(S_3^+ : S_t)$, consider the following lower bound example. Given an integer k, let $G = A_1 \cup A_2 \cup A_3 \cup H$ where H is a rainbow triangle-free coloring of a complete graph on $k - 2$ vertices where using colors $4, \ldots, k$, and A_i is a complete graph of order $\frac{n-k+2}{3}$ colored entirely with color i for each $i = 1, 2, 3$. The edges of $E(A_1, A_2)$ have color 3, $E(A_2, A_3)$ have color 1 and $E(A_1, A_3)$ have color 2. Also all edges of $E(H, A_1)$ have color 3, $E(H, A_2)$ have color 1 and edges of $E(H, A_3)$ have color 2. The graph G contains no rainbow S_3^+ but contains a monochromatic star of order $\frac{n+2k-4}{3}$. In light of this example, the authors of [22] conjectured the following.

Conjecture 8 ([22]) For all $k \geq 4$,

$$gr'_k(S_3^+ : S_t) = 3t - 2k + 4_1$$

References

1. R.J. Faudree, R. Gould, M. Jacobson, C. Magnant, Ramsey numbers in rainbow triangle free colorings. Australas. J. Combin. **46**, 269–284 (2010)
2. C. Magnant, Z. Magnant, K. Ozeki, *On Gallai-Ramsey Numbers for Paths*. Manuscript
3. S. Fujita, C. Magnant, Gallai-Ramsey numbers for cycles. Discrete Math. **311**(13), 1247–1254 (2011)
4. D. Bruce, Z. Song, Gallai-Ramsey numbers of C_7 with multiple colors. Discrete Math. **342**(4), 1191–1194 (2019)
5. J. Gregory, C. Magnant, Z. Magnant, *Gallai-Ramsey Number for the* 8-Cycle. Submitted
6. C. Bosse, Z. Song, *Multicolor Gallai-Ramsey Numbers of C_9 and C_{11}*. Submitted
7. C. Bosse, Z. Song, J. Zhang, *Multicolor Gallai-Ramsey Numbers of c_{13} and c_{15}*. Submitted
8. M. Axenovich, P. Iverson, Edge-colorings avoiding rainbow and monochromatic subgraphs. Discrete Math. **308**(20), 4710–4723 (2008)
9. F.R.K. Chung, R.L. Graham, Edge-colored complete graphs with precisely colored subgraphs. Combinatorica **3**(3–4), 315–324 (1983)
10. A. Gyárfás, G. Sárközy, A. Sebő, S. Selkow, Ramsey-type results for gallai colorings. J. Graph Theory **64**(3), 233–243 (2010)
11. H. Liu, C. Magnant, A. Saito, I. Schiermeyer, Y. Shi, *Gallai-Ramsey Number for K_4* (Accepted - J, Graph Theory, 2019)
12. C. Magnant, I., Schiermeyer, *Gallai-Ramsey Number for K_5*. Submitted
13. H. Wu, C. Magnant, P. Salehi Nowbandegani, S. Xia, All partitions have small parts—Gallai-Ramsey numbers of bipartite graphs. Discrete Appl. Math. **254**, 196–203 (2019)
14. H. Wu, C. Magnant, Gallai-Ramsey numbers for monochromatic triangles or 4-cycles. Graphs Combin. **34**(6), 1315–1324 (2018)
15. J. Zou, Y. Mao, C. Magnant, Z. Wang, C. Ye, Gallai-Ramsey numbers for books. Discrete Appl. Math. **268**, 164–177 (2019)
16. A. Gyárfás, G. Simonyi, Edge colorings of complete graphs without tricolored triangles. J. Graph Theory **46**(3), 211–216 (2004)
17. Z. Wang, C. Magnant, Y. Mao, J. Zou, *Ramsey and Gallai-Ramsey Numbers for Two Classes of Unicyclic Graphs*. Submitted
18. Z. Wang, Y. Mao, C. Magnant, I. Schiermeyer, J. Zou, *Gallai-Ramsey Numbers of Odd Cycles*. Submitted
19. F. Zhang, Y. Chen, Z. Song, *Gallai-Ramsey Numbers of Cycles*. Submitted

20. J. Fox, A. Grinshpun, J. Pach, The Erdős-Hajnal conjecture for rainbow triangles. J. Combin. Theory Ser. B **111**, 75–125 (2015)
21. C. Magnant, A general lower bound on Gallai-Ramsey numbers for non-bipartite graphs. Theo. Appl. Graphs **5**(1), Article 4 (2018)
22. S. Fujita, C. Magnant, Extensions of Gallai-Ramsey results. J. Graph Theory **70**(4), 404–426 (2012)

Index

A
Adjacent, 1

B
Bipartite, 3
Block, 4
Blow-up, 26
Book, 72
Bridge, 4
Bristles, 32
Broom, 32

C
Chromatic number, 26
Clique, 2
Color degree, 4
Colored, 1
Complement, 4
Complete bipartite, 3
Complete graph, 2
Component, 4
Connected, 4
Copy, 2
Cut edge, 4
Cut vertex, 4
Cycle, 3

D
Degree, 2
Disconnected, 4

E
Edge-colored, 1

Empty, 4
End vertex, 2

F
Finite, 1

G
Gallai coloring, 5, 9
Gallai partition, 10
Gallai-Ramsey number, 5
Graph, 1

H
Handle, 32

I
Independence number, 4
Independent, 4
Induced, 1
Isolated vertex, 2
Isomorphic, 1

J
Join, 3

K
k-connected, 4
k-edge-connected, 4

L
Leaf, 2
Little-o notation, 4

C. Magnant and P. Salehi Nowbandegani, *Topics in Gallai-Ramsey Theory*,
SpringerBriefs in Mathematics, https://doi.org/10.1007/978-3-030-48897-0

Printed in the United States
By Bookmasters